高等学校"十三五"规划教材

有机化学微型实验

第三版

朱红军 刘 睿 主编

王锦堂 主审

化学工业出版社

·北京·

《有机化学微型实验》(第三版)根据有机化学发展趋势和有机化学教学改革动态,结合有机化学实验微型化的特点,针对性地充实微型化、绿色化的课程内容。全书共分3章,分别介绍了有机化学微型实验的基本要求、基本操作和注意事项,以及24大类85个实验。第三版除增加英文有机化学实验内容外,还更新了SciFinder Scholar、Reaxys和Web of Science等网络检索资源和软件的介绍说明,更好地适应了有机化学实验的教学要求和有利于科研人员参考。

《有机化学微型实验》(第三版)可作为高等院校化学类专业或相关专业本科生的教材,也可作为广大从事医药、农药、应用化学、日化、轻化工、生物化工、材料化学等领域科学研究的工程技术人员的参考书。

图书在版编目(CIP)数据

有机化学微型实验/朱红军,刘睿主编;王锦堂主审. —3版. —北京:化学工业出版社,2018.8(2023.1重印)
高等学校"十三五"规划教材
ISBN 978-7-122-32599-0

Ⅰ.①有⋯　Ⅱ.①朱⋯②刘⋯③王⋯　Ⅲ.①有机化学-化学实验-高等学校-教材　Ⅳ.①O62-33

中国版本图书馆CIP数据核字(2018)第149264号

责任编辑:宋林青　　　　　　　　　　　文字编辑:刘志茹
责任校对:王　静　　　　　　　　　　　装帧设计:关　飞

出版发行:化学工业出版社(北京市东城区青年湖南街13号　邮政编码100011)
印　　装:三河市延风印装有限公司
787mm×1092mm　1/16　印张12¾　字数307千字　2023年1月北京第3版第5次印刷

购书咨询:010-64518888　　售后服务:010-64518899
网　　址:http://www.cip.com.cn
凡购买本书,如有缺损质量问题,本社销售中心负责调换。

定　价:28.00元　　　　　　　　　　　　　　　　　　　版权所有　违者必究

前　言

有机化学微型实验（Microscale Organic Experiment）自20世纪80年代初在国际上崛起并取得成功之后，因其具有简化仪器、节约化学试剂、节约动力消耗、节省时间、减少环境污染、实验条件易于控制、有利于培养学生严谨的科学态度、降低实验员的劳动强度等常规实验无法比拟的优越性，在全球化学实验教学领域备受关注。本教材自2001年出版以来，为许多高校有机化学实验的改革提供了素材。广大学生、教师和科技工作者在使用过程中本着对本教材的爱护，提出了不少意见。在国际学术化交流日益广泛、学科间交叉渗透日益频繁的今天，对有机化学微型实验教材进行双语化改革探索，既顺应我国高等教育国际化趋势，也是学科与时俱进的发展需要。为此我们对本教材进行了再版修订。

在第二版的基础上，根据有机化学发展趋势和有机化学教学改革动态，结合有机化学实验微型化的特点，对传统实验教学内容进行了必要的取舍，并针对性地充实微型化、绿色化的课程内容。遵循前沿化、国际化、微型化、绿色化的原则，着重解决教学内容的先进性与实用性，除增加一些英文有机化学实验项目外，还更新了SciFinder Scholar、Reaxys和Web of Science等网络检索资源和软件的介绍说明，以更好地适应有机化学实验教学的要求和有关院校、企业和科研院所的科研人员进行精细有机合成研究的参考要求。

南京工业大学朱红军、刘睿、解沛忠、宋广亮、邹冲和上海理工大学李钰皓等老师参加了本书的编写与修订工作，南京工业大学化学与分子工程学院研究生陆佳鹏、胡锦阳也参与了本书的部分文字工作。

限于编者水平，书中难免有疏漏和不妥之处，希望各位读者给予批评指正。

编者
2018年4月8日

第一版前言

有机化学微型实验是20世纪80年代在国际上崛起，近年来在国内外迅速发展起来的一种新颖的实验方法和实验技术，与常量实验相比，它具有以下突出的优点。

（1）可节约实验的仪器费用和药品费用。微型实验仪器小，相对成本低，药品用量少，消耗少。仪器小，占空间少，大大提高了实验室利用率，缓解了实验室紧缺的困难。

（2）微型实验仪器小，药品用量少，实验精度增加，它促使学生必须集中精力，规范操作，方能获得预期结果，提高了对学生的要求。有利于培养学生严谨的科学态度、规范化的操作技能，有利于提高实验课教学质量。

（3）提高了学生的科研工作能力，避免了实验课中长期沿用的常量实验与科研工作中微量、半微量实验的脱节现象。

（4）微型实验反应物用量少，反应及产品后处理时间也相应减少，又加上仪器装拆简单、操作方便，所以可缩短实验时间。

（5）微型实验处理的药品量少，相应产生的废气、废水和废渣少，可大大减少对环境的污染。

因此，有机化学微型实验的开设具有显著的经济效益、环保效益和教学效果，代表了有机化学实验改革发展的方向，是传统有机化学实验的提高。本实验教材的主要目的是帮助学生通过进行有机化学微型实验，掌握有机化学实验的基本操作技能；在分析问题和解决问题方面得到更严格的训练，培养高素质人才，以更好地适应社会的要求，同时也节约一定的成本。本教材可作为高等院校有机化学实验的教材和参考书。

本教材既对原有的经典仪器进行了系统的介绍，同时也对新型的仪器作了介绍，如电子天平、数字式压力计、紧密大气压力计、电热套、磁力搅拌器等，这些仪器精确、使用方便，可以大大节约实验时间，代表了一种发展趋势，会逐步取代原有的仪器。

本教材也安排了有关药物的合成、农药的合成、光化学等方面的内容，因此本书同时也可作为"制药"、"应用化学"、"轻化工程"等专业课的实验教材，并且也可作为有关工矿企业和研究院所的科研人员进行精细有机合成研究的一本参考书。

实验指导教师、学生和研究人员在使用本教材时，应注意掌握有机化学微型实验仪器的正确使用方法、原理及其在有机合成实验中的应用，还要注意常量实验与微型实验两种类型仪器的异同。

南京工业大学应用化学系朱红军主编了本书，参加编写的还有汪海波、万嵘、殷峻、张荣。王锦堂教授对全书进行了审阅。

由于编者水平的限制和时间仓促等方面的问题，本书还存在着很多需要改进的地方。希望各位读者对本书的不足之处给予指正。

第二版前言

本教材自 2001 年出版以来受到广大读者的热爱，为许多高校有机化学实验的改革提供了素材。广大学生、教师和科技工作者在使用过程中本着对该教材的爱护，提出了不少意见，同时有机化学实验新技术也不断出现，为此我们对本教材进行了修订。

本教材第二版在第一版的基础上增加了 SciFinder、ChemFinder 和 Beilstein 等网络检索资源和软件，样品干燥、常量实验仪器的介绍、微波和超声波等非常规合成技术等，使本教材更具有先进性和实用价值，更好地适应有机化学实验教学的要求和有关工矿企业和研究院所的科研人员进行精细有机合成研究参考的要求。

南京工业大学朱红军教授和李玉峰等老师对本教材的修订进行了有关工作。

限于编者水平，本书还存在很多值得改进的地方，希望各位读者对本书的不足之处给予指正。

编　者
2007 年 2 月

目　录

1 绪论 ……………………………………………………………………………………… 1
　1.1 有机化学实验基本要求 ………………………………………………………………… 1
　1.2 有机化学实验注意事项 ………………………………………………………………… 1
　　1.2.1 安全注意事项 …………………………………………………………………… 1
　　1.2.2 有机化学实验预习要求 ………………………………………………………… 2
　　1.2.3 有机化学实验要求 ……………………………………………………………… 2
　　1.2.4 有机化学实验报告格式 ………………………………………………………… 2
　1.3 有机化学微型实验常用仪器简介 ……………………………………………………… 4
　　1.3.1 玻璃仪器 ………………………………………………………………………… 4
　　1.3.2 常用仪器设备 …………………………………………………………………… 7
　　1.3.3 微型实验药品的称量 …………………………………………………………… 12
　1.4 手册的查阅及有机化学文献简介 ……………………………………………………… 12
　　1.4.1 工具书和参考书 ………………………………………………………………… 12
　　1.4.2 期刊 ……………………………………………………………………………… 13
　　1.4.3 化学文摘 ………………………………………………………………………… 13
　　1.4.4 网上资源 ………………………………………………………………………… 13

2 基本操作与注意事项 …………………………………………………………………… 20
　2.1 蒸馏和分馏 …………………………………………………………………………… 20
　　2.1.1 蒸馏和分馏装置 ………………………………………………………………… 20
　　2.1.2 蒸馏和分馏操作 ………………………………………………………………… 20
　2.2 减压蒸馏 ……………………………………………………………………………… 21
　　2.2.1 减压蒸馏装置 …………………………………………………………………… 23
　　2.2.2 减压蒸馏操作 …………………………………………………………………… 23
　2.3 水蒸气蒸馏 …………………………………………………………………………… 27
　2.4 熔点测定 ……………………………………………………………………………… 28
　　2.4.1 毛细管法测定熔点 ……………………………………………………………… 29
　　2.4.2 显微熔点测定仪测定熔点 ……………………………………………………… 30
　　2.4.3 Melting Point Determination of Benzoic Acid ……………………………… 32
　2.5 重结晶 ………………………………………………………………………………… 33
　　2.5.1 重结晶过程 ……………………………………………………………………… 33
　　2.5.2 溶剂的选择及用量 ……………………………………………………………… 33
　　2.5.3 溶解及热过滤 …………………………………………………………………… 35

- 2.5.4 结晶、抽滤和干燥 ·· 36
- 2.5.5 Recrystallization of Benzoic Acid ································· 38
- 2.6 折射率的测定 ·· 40
- 2.7 萃取 ·· 42
  - 2.7.1 分次萃取法 ·· 43
  - 2.7.2 连续萃取法 ·· 43
- 2.8 干燥 ·· 44
  - 2.8.1 液体的干燥 ·· 44
  - 2.8.2 固体的干燥 ·· 45
- 2.9 升华 ·· 47
  - 2.9.1 萘的常压升华 ··· 48
  - 2.9.2 萘的减压升华 ··· 48
- 2.10 色谱 ··· 48
  - 2.10.1 柱色谱 ··· 48
  - 2.10.2 薄层色谱 ··· 50
  - 2.10.3 气相色谱 ··· 51
  - 2.10.4 高效液相色谱 ··· 52
- 2.11 波谱技术 ··· 53
  - 2.11.1 红外光谱（IR） ·· 53
  - 2.11.2 紫外光谱（UV） ··· 54
  - 2.11.3 核磁共振谱（NMR） ··· 55
  - 2.11.4 质谱（MS） ··· 57
- 2.12 Schlenk Technique and Glove Box Operation ······················· 58
  - 2.12.1 Schlenk Technique ··· 58
  - 2.12.2 Glove Box Operation ·· 60
- 2.13 Reaction Monitoring ·· 62
  - 2.13.1 Thin Layer Chromatography（TLC） ························· 62
  - 2.13.2 High Performance Liquid Chromatography（HPLC） ····· 67
  - 2.13.3 Gas-liquid Chromatography（GC，GLC，VPC） ········· 70
- 2.14 Some Methods of Separation ··· 73
  - 2.14.1 Azeotropic and Extractive Distillation ·························· 73
  - 2.14.2 Supercritical Fluid Extraction ····································· 74
  - 2.14.3 Liquid-Liquid Systems ··· 75
  - 2.14.4 Leaching ·· 75

3 实验 ·· 77
- 3.1 烯烃的制备 ··· 77
  - 实验1 环己烯 ·· 77
  - Experiment 2 Synthesis of 2-Methyl-2-butene ························· 78
- 3.2 卤代烃的制备 ·· 79

  实验 3　溴乙烷 ··· 79

  实验 4　溴丁烷 ··· 80

  实验 5　α,β-二溴苯乙烷 ·· 81

  Experiment 6　Synthesis of *t*-Butyl Chloride ··· 82

 3.3　醇的制备 ·· 84

  实验 7　苯甲醇 ··· 84

  实验 8　二苯甲醇 ·· 85

  Experiment 9　Synthesis of Triphenylmethanol ·· 86

 3.4　醚的制备 ·· 88

  实验 10　乙醚 ··· 88

  实验 11　正丁醚 ·· 90

  Experiment 12　Synthesis of 4-Nitroanisol ·· 91

 3.5　醛酮的制备 ··· 92

  实验 13　环己酮 ·· 92

  实验 14　苯乙酮 ·· 93

  实验 15　茴香基丙酮 ··· 94

  Experiment 16　Synthesis of Benzophenone ·· 95

  Experiment 17　Synthesis of Diphenylethanedione（Benzil） ··························· 97

 3.6　羧酸的制备 ··· 98

  实验 18　苯甲酸 ·· 98

  实验 19　对硝基苯甲酸 ··· 99

  实验 20　己二酸 ··· 100

  实验 21　香豆素-3-羧酸 ··· 101

  Experiment 22　Synthesis of Cinnamic Acid ·· 102

 3.7　羧酸衍生物 ··· 103

  实验 23　邻苯二甲酸二丁酯 ·· 103

  实验 24　丁二酸酐 ··· 104

  实验 25　苯甲酸乙酯 ··· 105

  实验 26　巯基乙酸铵 ··· 106

  Experiment 27　Synthesis of Ethyl Acetate ·· 107

  Experiment 28　Synthesis of Acetanilide ··· 109

 3.8　芳香族硝基化合物 ·· 110

  实验 29　硝基苯 ··· 110

  实验 30　间二硝基苯 ··· 111

  Experiment 31　Synthesis of 2-Nitroresorcinol ·· 112

 3.9　胺及季铵盐 ··· 114

  实验 32　间硝基苯胺 ··· 114

  实验 33　邻氨基苯甲酸 ··· 115

  实验 34　溴化四丁基铵 ··· 115

  实验 35　溴化四乙基铵 ··· 116

Experiment 36　Synthesis of Benzyltriethylammonium Chloride ……………… 116
3.10　重氮盐及其反应 …………………………………………………………………… 117
　　实验 37　氯苯 …………………………………………………………………… 117
　　实验 38　甲基红 ………………………………………………………………… 119
　　实验 39　Ⅱ号橙 ………………………………………………………………… 119
　　Experiment 40　Synthesis of Methyl Orange …………………………………… 120
3.11　芳香族磺酸 ………………………………………………………………………… 121
　　实验 41　双酚-S ………………………………………………………………… 121
　　Experiment 42　Synthesis of p-Toluene Sulfonic Acid ………………………… 123
3.12　缩合反应 …………………………………………………………………………… 124
　　实验 43　双酚 A ………………………………………………………………… 124
　　实验 44　安息香 ………………………………………………………………… 124
　　Experiment 45　Synthesis of Ethyl Acetoacetate ………………………………… 126
3.13　Diels-Alder 反应 …………………………………………………………………… 128
　　实验 46　9,10-二氢蒽-9,10-α,β-马来酸酐 …………………………………… 128
3.14　坎尼扎罗反应 ……………………………………………………………………… 128
　　实验 47　呋喃甲醇和呋喃甲酸 ………………………………………………… 128
　　Experiment 48　Synthesis of Benzoic Alcohol and Benzoic Acid ……………… 130
3.15　格利雅反应 ………………………………………………………………………… 131
　　实验 49　苯甲酸 ………………………………………………………………… 131
　　实验 50　2-甲基-2-己醇 ………………………………………………………… 133
　　Experiment 51　Synthesis of Malachite Green and Crystal Violet ……………… 134
3.16　催化氢化反应 ……………………………………………………………………… 135
　　实验 52　3-苯基丙酸(氢化肉桂酸) …………………………………………… 135
　　Experiment 53　Synthesis of cis-4-t-Butylcyclohexanol ……………………… 137
　　Experiment 54　Reduction of Camphor ………………………………………… 137
3.17　天然产物的提取和分离 …………………………………………………………… 139
　　实验 55　绿色植物色素的提取 ………………………………………………… 139
　　实验 56　从西红柿中提取番茄红素和 β-胡萝卜素 ………………………… 140
　　实验 57　果胶的提取 …………………………………………………………… 142
　　Experiment 58　Extraction of Caffeine from Tea ………………………………… 143
3.18　光学异构体的拆分 ………………………………………………………………… 145
　　实验 59　苯乙胺的制备及外消旋体的拆分 …………………………………… 145
3.19　相转移催化和卡宾反应 …………………………………………………………… 147
　　实验 60　扁桃酸 ………………………………………………………………… 147
　　实验 61　1-苯基-2,2-二氯环丙烷 ……………………………………………… 148
　　Experiment 62　Synthesis of 7,7-Dichlorobicyclo [4.1.0] heptane ……………… 149
3.20　碳水化合物及其衍生物 …………………………………………………………… 150
　　实验 63　五乙酸葡萄糖酯 ……………………………………………………… 150
　　实验 64　从牛奶中制取酪蛋白和乳糖 ………………………………………… 151

实验 65　食品增稠剂、保型剂乙酯淀粉的制备 ……………………………………………… 152
3.21　氨基酸与肽 ………………………………………………………………………………… 153
　　实验 66　L-酪氨酸甲酯（L-Tyrosine methyl ester） ……………………………………… 153
3.22　非常规有机合成方法 ……………………………………………………………………… 153
　　实验 67　化学发光指示剂 3-氨基邻苯二甲酰肼 …………………………………………… 153
　　实验 68　二苯甲酮的光化学还原 …………………………………………………………… 155
　　实验 69　微波辐射下的 Perkin 反应——肉桂酸的制备 …………………………………… 156
　　实验 70　微波辐射下苯并吡喃-2-酮-3-甲酸乙酯的制备 …………………………………… 156
　　实验 71　超声条件下 3-烯丙基-2,4-戊二酮的制备 ………………………………………… 157
　　Experiment 72　Synthesis of Methyl Benzoate under Ultrasonic Radiation ……………… 158
　　Experiment 73　Synthesis of 1,2,3,4-Tetrahydrocarbazole ………………………………… 159
3.23　金属有机化合物 …………………………………………………………………………… 160
　　实验 74　二茂铁的合成 ……………………………………………………………………… 160
　　实验 75　丁基锂的合成 ……………………………………………………………………… 162
　　Experiment 76　Synthesis of Acetyl Ferrocene …………………………………………… 163
3.24　药物的合成 ………………………………………………………………………………… 165
　　实验 77　局部麻醉剂苯佐卡因 ……………………………………………………………… 165
　　实验 78　磺胺药物——磺胺 ………………………………………………………………… 166
　　实验 79　解热镇痛药——非那西汀 ………………………………………………………… 167
　　实验 80　止咳酮 ……………………………………………………………………………… 168
　　实验 81　抗癫痫药 5,5-二苯基乙内酰脲 …………………………………………………… 169
　　Experiment 82　Synthesis of Aspirin（acetylsalicylic acid） ……………………………… 170
3.25　农药的合成 ………………………………………………………………………………… 172
　　实验 83　杀灭菊酯（氰戊菊酯） …………………………………………………………… 172
　　实验 84　O,O-二甲基-O-(2,2-二氯乙烯基)磷酸酯——敌敌畏 ……………………… 173
　　Experiment 85　Synthesis of Anophelifuge N,N-Diethyl-m-toluamide ……………… 173

附录 …………………………………………………………………………………………………… 175
　　附录 1　常用有机试剂的物理性质 …………………………………………………………… 175
　　附录 2　常用的酸和碱 ………………………………………………………………………… 176
　　附录 3　常用酸碱溶液密度组成表 …………………………………………………………… 177
　　附录 4　常用有机溶剂处理 …………………………………………………………………… 181
　　附录 5　危险化学品使用知识 ………………………………………………………………… 184
　　附录 6　典型的红外光谱吸收谱带 …………………………………………………………… 187
　　附录 7　一些常见基团质子的化学位移值 …………………………………………………… 189
　　附录 8　实验单元操作一览表 ………………………………………………………………… 189

参考文献 ……………………………………………………………………………………………… 192

1 绪 论

1.1 有机化学实验基本要求

有机化学实验是有机化学教学的重要组成部分，目的是培养学生的基本操作技能，进行理论、原理运用到实践的训练，培养学生发现问题、分析问题和解决问题的能力，养成严肃认真、实事求是的科学态度和严谨的工作作风，从而使学生在科学方法上得到初步的训练，为今后在实际工作中进行科学研究、生产管理等奠定良好的基础。

有机化学实验要求学生掌握玻璃仪器的清洗、干燥、装配和拆卸，加热和冷却，回流、蒸馏、分馏、减压蒸馏、萃取和洗涤，重结晶和过滤，液体和固体样品的干燥，沸点、熔点和折射率的测定等基本操作技能和文献检索技能；以及将这些基本操作技能运用到有机化合物的合成路线设计、方案确立与实施、分离提纯和物理常数的测定、结构鉴定等方面。

1.2 有机化学实验注意事项

1.2.1 安全注意事项

干任何事情，安全第一，有机化学实验尤其如此。这是因为有机化合物易燃、易爆、毒性高，有机反应的副反应多，难以控制。在有机化学实验过程中，在安全方面需要注意以下事项：

(1) 思想上要重视，不能松懈。
(2) 预习要充分，注意了解原材料和产物的性能，以及反应在安全方面的问题，做到心中有数。
(3) 实验者进入实验室，首先要了解、熟悉实验室电闸、煤气开关、水源开关及消防器材如灭火器、沙箱、石棉布等的放置地点，不得随意移动消防器材的位置。
(4) 实验时严格遵守实验室安全守则和实验室规章制度，穿好实验服、佩戴胶皮手套及防护眼镜等防护用品，严格按实验操作要求认真操作，不能马虎。
(5) 出现事故时，首先要冷静，正确处理，及时报告。
(6) 掌握常见的事故处理方法和常用消防器材的使用方法。

常见事故的预防和处理方法：

(1) 实验开始前，要仔细检查仪器有无破损、装置是否正确稳妥。
(2) 易燃溶剂在操作时应该远离明火，不能敞口加热。
(3) 在操作易挥发的液体、有毒气体时，必须在通风橱内进行，同时对有毒气体出口处应该进行吸收处理。
(4) 发生着火事故时，首先要冷静，及时关闭气源、电源，并报告指导教师，对小火可用湿抹布或石棉布覆盖。对大面积着火，应该及时报警，并用二氧化碳灭火器、干粉灭火器或四氯化碳灭火器灭火。如果衣服着火，应立即就近使用喷淋器或立即在地面上打滚，将火熄灭。或将衣服迅速脱下扑灭，千万不能惊慌乱跑，以避免使火焰扩大和烧向头部。

（5）如果发生腐蚀性药品溅到皮肤上或眼睛里，一般应该立即用大量的水进行清洗。如果是浓酸引起的，可用3％碳酸氢钠溶液清洗；如果是浓碱引起的，则用1％的硼酸溶液清洗。最后再水洗。对比较严重的情况初步处理后应该立即到医院进行治疗。

（6）如遇割伤，首先将伤口的玻璃等碎渣清除干净，再用水和碘酒清洗伤口，包扎好伤口后到医院进行治疗。

1.2.2 有机化学实验预习要求

进行有机化学实验必须预习，没有预习，不能进行有机化学实验操作。充分、正确的预习是保证实验安全的前提之一，也是掌握实验技能、提高实验效率必不可少的一步。

实验预习时应该了解实验目的、实验原理、原材料及其物理常数、仪器名称及其使用方法和装置搭接方法、实验操作步骤及其注意事项。

预习方法和途径：预习实验教材和教辅书的有关内容；查阅有机化学教材中有关内容的基本原理；检索有关内容研究进展和有关背景知识；查阅有关原材料与产品等的物理常数；也可到相关网站上查阅有关实验的介绍和通过有关的测试与考核。

实验预习时应做好预习报告，报告应该包含以上内容。没有预习报告，不得进入实验室进行实验操作。

1.2.3 有机化学实验要求

（1）认真听指导教师的讲解，服从指导教师及实验员的安排。

（2）实验服、防护手套及眼镜穿戴齐全，严格按照实验操作步骤进行实验。

（3）仔细观察实验现象，并实事求是地记录实验数据及实验现象。

（4）实验时要开动脑筋，积极思考，不盲目操作。

（5）实验过程中不得随便离开实验场所，不得大声喧哗，不得看与实验无关的书籍、报纸等。

（6）实验过程中出现的异常现象应该及时报告给指导教师。

（7）实验时注意安排好前后次序，合理安排时间，做到既快又好，养成良好的工作作风。

（8）要爱护实验仪器、设备和环境。

（9）实验结束后，玻璃仪器要清洗干净，打扫卫生，把水、电、气、火和门窗关好。

（10）有关事项完成后，报告指导教师方能离开实验室。

1.2.4 有机化学实验报告格式

实验报告是对一个实验的全面总结，它包括以下部分：

（1）实验目的与要求。

（2）实验原理，包括主反应式与副反应式等。

（3）实验仪器与装置图。

（4）药品规格与用量。

（5）原材料与产品及副产物的物理常数。

（6）实验操作步骤与现象记录。

（7）实验结果。

（8）实验结果讨论。

有机化学实验英文报告的内容与中文实验报告类似，参考格式如下：

Experiment Report

Name		Class		Student ID	
Laboratory Temperature		Laboratory Pressure		Date	
Title of Experiment					

1. Experimental Objectives

2. Experimental Principles (Chemical Reactions)

3. Apparatus and Reagents

4. Apparatus Illustrations

5. Key Operation Points

6. Experimental Record			
Time	Operation	Data and Observation	Notes

7. Results and Discussion

1.3 有机化学微型实验常用仪器简介

1.3.1 玻璃仪器
1.3.1.1 玻璃仪器名称

根据微型化学实验的特点，常见有机化学微型实验制备仪如图 1-1 所示（有机化学常量仪器也包括在其中）。

单口圆底烧瓶
round bottom flask

二口圆底烧瓶
two-necked round bottom flask

三口圆底烧瓶
three-necked round bottom flask

烧杯
beaker

茄形瓶
eggplant-shaped flask

锥形瓶
conical flask

热滤漏斗
hot filtration funnel

短颈漏斗
short stem funnel

布氏漏斗
Buchner funnel

磨口三角漏斗
powder funnel

直形冷凝管
Liebig condenser

球形冷凝管
Allihn condenser

空气冷凝管
air reflux condenser

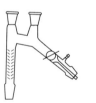
H形分馏头
H-type distillation adapter

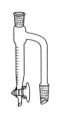
分水器
Dean-Stark trap

蒸馏头
distillation adapter

韦氏蒸馏头
Vigreux distillation adapter

索氏抽提器
Soxhlet extractor

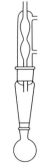
固液提取器
solid-liquid extractor

分液漏斗
separatory funnel

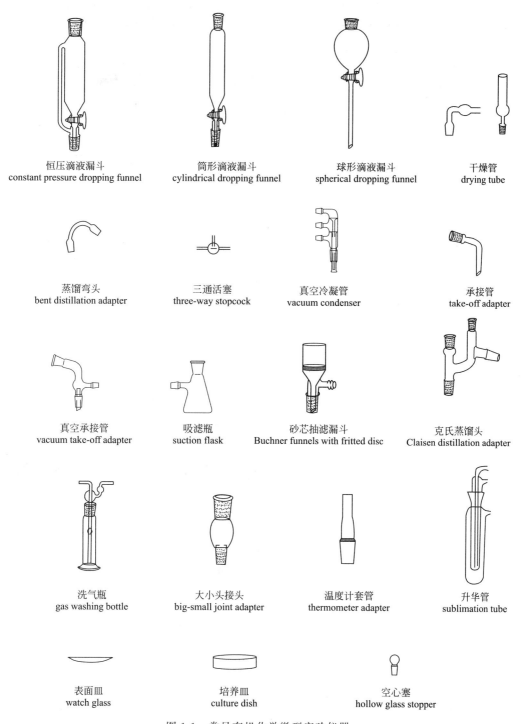

图 1-1 常见有机化学微型实验仪器

Figure 1-1 Common apparatuses for microscale organic chemistry experiments

本套仪器用于有机化学微型实验的教学和科研中，具有装拆简单、使用方便、操作规范准确等特点。特别是 H 形分馏头，用于常、减压蒸（分）馏时，能准确测量各馏分温度，可以接收多馏分，能使单步及多步蒸、分馏操作连续进行。

利用本套仪器可以组合成反应装置（如图1-2～图1-6所示）、分液装置、抽滤装置、升华装置、常压蒸（分）馏装置、减压蒸（分）馏装置、水蒸气蒸馏装置、固/液提取装置等有机化学实验中常用的装置。

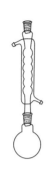  

(a) 普通回流反应装置　　　　　　(b) 干燥回流反应装置　　　　　(c) 气体吸收反应装置
(a) normal reflux reaction apparatus　(b) dry and reflux reaction apparatus　(c) gas absorption reaction apparatus

图 1-2　回流反应装置

Figure 1-2　Reflux reaction apparatuses

 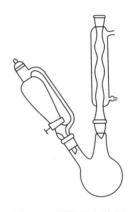 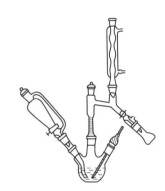

图 1-3　分水反应装置　　　　　图 1-4　滴液反应装置　　　　　图 1-5　其他反应装置

Figure 1-3　Water segregator apparatus　　Figure 1-4　Dropping reaction apparatus　　Figure 1-5　Other reaction apparatus

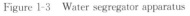

 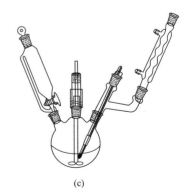

(a)　　　　　　　　　(b)　　　　　　　　　(c)

图 1-6　搅拌装置

Figure 1-6　Stirring apparatuses

1.3.1.2 仪器的清洗

在进行实验时,为避免杂质进入反应物中,必须用清洁的玻璃仪器盛装。洗涤玻璃仪器最简单的方法是用毛刷和湿润的去污粉擦洗,再用清水冲净,将仪器倒置。

微型玻璃仪器中的圆底烧瓶、茄形瓶、锥形瓶、漏斗、直形(球形)冷凝管等均可用小试管刷和去污粉来擦洗。擦洗时可根据容器具体形状将试管刷弯曲后使用。若有固体物黏附在器壁上,可用小刮刀刮下来,然后再用刷子擦洗。若烧瓶用水无法洗净,可用回收溶剂擦洗,也可用洗液清洗,必要时可适当加热,但要注意防火。对于三角滤板漏斗,可先用适宜的回收溶剂抽洗,再用清水冲净。清洗 H 形分馏头时,可装上烧瓶,用回收的溶剂(或直接用少量乙醚或乙醇)反复冲洗,特别是残留在器壁上的固体物,务必冲洗干净,随后再用清水冲洗残留的溶剂。切不可将毛刷伸进分馏柱内进行擦洗,以免损坏填料。所有仪器的磨口都要擦洗干净。仪器外壁可先用润湿的去污粉擦洗,然后用清水冲净。总之,实验前后都要保持仪器的清洁。

1.3.1.3 仪器的干燥

在有机化学实验中,许多反应都要求在无水条件下进行。因此,必须使用干燥的玻璃仪器。干燥仪器的方法有以下几种。

(1) 晾干。将洗净的仪器倒置,在空气中晾干。

(2) 在烘箱中保持 110~120℃,烘 0.5h,然后冷却到 60℃左右取出备用。仪器放入烘箱前应尽量把水甩干。对于带有活塞的仪器,如 H 形分馏头、多功能梨形漏斗、分水器等应取下活塞,擦净凡士林或真空脂,然后烘干。

(3) 热风吹干,用电吹风按冷风→热风→冷风的顺序吹干,或用吹风干燥仪器吹干。

(4) 用有机溶剂干燥。对急用的仪器,在用水洗净后先用少量乙醇冲洗几次,再用少量乙醚冲洗,然后用电吹风按冷风→热风→冷风的顺序吹干后即可使用。

1.3.1.4 仪器的使用方法

有机化学实验中的许多玻璃仪器为标准磨口仪器,价格较高,容易损坏,使用时需注意下述几个方面:

(1) 轻拿轻放,仔细操作,不能粗心大意;

(2) 对磨口处注意保护,防止固体杂物及碱性物质等对其的损伤;

(3) 搭建装置按从下至上,从左到右(或从右到左)的顺序进行,在固定之前应该转动,使磨口之间紧密接触,以防止泄漏,所搭建的仪器保持垂直,不能有扭力存在;

(4) 拆卸时,次序与装配时相反。

1.3.2 常用仪器设备

1.3.2.1 电子天平

电子天平是随着电子技术的发展而发展起来的一种新型称量工具,是电光天平的更新换代产品,具有操作方便、称量范围大、精确、稳定性好等特点。下面以 FA/JA 系列上皿电子天平为例来介绍电子天平的一般用法和注意事项。

该天平是采用 MCS-51 系列单片微机的多功能电子天平。除一般智能化电子天平所具有的称量自动校准,积分时间可调和灵敏度可适当选择外,本天平还有三种量制——克、米制克拉、金盎司供用户自由选择(米制克拉、金盎司供出口用)。数据接口配有 RS-232C 通用串行双向口和标准的 Centronics 并行输出口,能与微机和各种打印机(如 PP40 等)相连。并行数据输出还设有两种模式:定时和不定时输出。其中定时可设三挡不同的定时时间供用户自由选择。其面板和结构示意如图 1-7 所示。本天平采用轻触按键,能实行多键盘控制,操作灵活方便,各功能的转换与选择,只需按相应的按键。

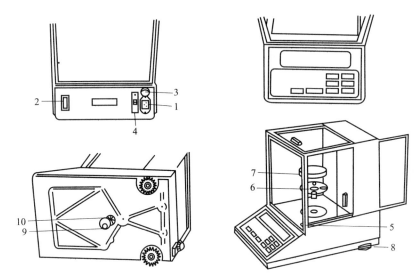

图 1-7 FA/JA 系列上皿式电子天平面板和结构示意图

1—电源插座;2—数据接口;3—保险丝;4—220V/110V 转换开关;5—水平仪;6—盘托;
7—秤盘;8—水平调节脚;9—盖板;10—挂钩

【开机】

· 在使用前观察水平仪。如水平仪水泡偏移,需调整水平调节脚,使水泡位于水平仪正中。

· 选择合适电源电压,将电压转换开关置于相应位置。

· 天平接通电源,就开始通电工作(显示器未工作),通常需预热 1h 后方可开启显示器进行操作使用(有关键盘的操作功能见附注)。

【称量】

按照附注选择模式(用户选定后,本天平由于具有记忆功能,所有选定模式能保持断电后不丢失就可用于称量)。

按 TAR 键,显示为零后,置被称物于秤盘上,待数字稳定(即显示器左边的"0"标志熄灭)后,所显示数字即为被称物的质量值。

【去皮重】

置容器于秤盘上,天平显示为容器质量值,按 TAR 键,显示零,即去皮重。再置被称物于容器中,这时显示的数字是被称物的净重。

【累计称量】

用去皮重称量法,将被称物逐个置于秤盘上,并相应逐一去皮清零,最后移去所有被称物,显示数的绝对值为被称物的总质量值。

【加物】

按住 INT 键不放,可调至 INT-0 模式,置容器于秤盘上,去皮重。将被称物(液体或松散物）逐步加入容器中，能快速得到连续读数值。当加物达到所需称量，显示最左边"0"熄灭，这时显示的数值即为用户所需的称量值。当加入混合物时，可用去皮重法，对每种物质计净重。

【读取偏差】

置基准砝码（或样品）于秤盘上，去皮重，然后取下基准码，显示其质量负值。再置称物于秤盘上，视称物比基准砝码重或轻，相应显示正或负偏差值。

【下称】

拧松天平底部下盖板的螺丝，露出挂钩。将天平置于开孔的工作台上，调整水平，并对天平进行校准工作，就可用挂钩称量挂物了。

【天平的维护与保养】

天平必须小心使用。秤盘与外壳须经常用软布和牙膏轻轻擦洗。切不可用强溶解剂擦洗。

【附注】

键盘的操作功能

① 开启显示器键（ON）

只要轻按一下 ON 键，显示器全亮，如下所示。

$$\pm_0 8888888\,^\%_g$$

对显示器的功能进行检查，约 2s 后显示天平的型号，然后是称量模式。

| 0.0000g | 或 | 0000g |

② 关闭显示器键（OFF）

轻按 OFF 键，显示器熄灭即可。若要较长时间不使用天平，应拔去电源线。

③ 清零、去皮键（TAR）

置容器于秤盘上，显示出容器的质量。然后按 TAR 键，显示消失，随即出现全零状态，表示容器质量显示值已去除，即去皮重。当移去容器，就出现容器质量的负值。再轻按 TAR 键，显示器为全零，即天平清零。

④ 称量范围转换键（RNG）

本天平具有 2 种读数精度。称量范围在 0～30g，其读数精度为 ±0.1mg。若总质量超过 30g，天平就自动转为 ±1mg 读数精度。但通过具有 0～160g 的去除皮重功能，在总质量不超过 160g 的范围内可分段（其分析量不超过 30g）进行读数达 ±0.1mg 精度的分析。即若容器总量超过 30g，可轻按 TAR 键，先去除容器质量，然后称物（≤30g），其显示读数精度仍为 ±0.1mg。

称量范围设置：

只要按住 RNG 键不松手，显示器就会如下所示，不断循环。

| mg—30 | ⇌ | mg—160 |

如需要读数精度为 ±0.1mg 时，当显示器出现 mg—30 即松手，随即出现等待状态————，最后出现称量状态。

⑤ 量制转换键（UNT）

按住 ERY 键不松手，显示器就会显现如下，不断循环。

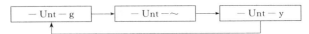

"g"表示单位克，"～"表示"米制克拉"，"y"表示单位为金盎司。量制单位的内体设置同 RNG 键。

⑥ 积分时间调整键

积分时间有 4 个依次循环的模式可供选择。如下所示。

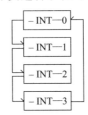

对应的积分时间长短为：

INT—0 快速；INT—1 短；INT—2 较短；INT—3 较长。

积分时间选定办法也与 RNG 键相同。

⑦ 灵敏度调整键（ASD）

同积分时间调整键一样，灵敏度也有依次循环的 4 种模式。

灵敏度顺序依次为：ASD—0 最高；ASD—1 高；ASD—2 较高；ASD—3 低。

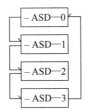

其中 ASD—0 是生产调试时用，用户不宜使用此模式，灵敏度模式的选定办法也与 RNG 键相同。

现将 ASD 与 INT 两模式的配合使用情况列出，以供用户参考。

最快称量速度　　　INT—1　ASD—3；
通常情况　　　　　INT—3　ASD—2；
环境不理想时　　　INT—3　ASD—3。

⑧ 天平校准键

因存放时间较长、位置移动、环境变化或为取得精确测量，天平在使用前一般都应进行校准操作。

校准天平的准备：取下秤盘上所有被称物，置 mg—30、INT—3、ASD—2、Ery—g 模式。轻按 TAR 键，天平清零。

校准天平：轻按 CAL 键，当显示器出现 CAL—时松手，显示器就出现 CAL—100，其中"100"为闪烁码，表示校准砝码需用 100g 的标准砝码。此时就把准备好的 100g 校准砝码放到秤盘上，显示器即出现"— — — —"等待状态，经较长时间后显示器出现 100.000g，拿去校准砝码，显示器应出现 0.000g，若出现不为零时，则再清零，再重复以上校准操作（注意：为了得到准确的校准结果，最好反复以上校准操作 2 次）。校准显示顺序如下。

⑨ 输出模式设定键（PRT）

按住 PRT 键也会有如下所示的 4 种模式依次循环出现，供用户随意选择。

PRT—0 为非定时按键输出模式。此时只要轻按一下 PRT 键，输出接口上就输出当时的称量结果 1 次。注意：此时应又轻又快地按此键，否则会出现下一个输出模式。

PRT—1 为定时 0.5min 输出 1 次；PRT—2 为定时 1min 输出 1 次；PRT—3 为定时 2min 输出 1 次。

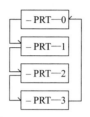

PRT 模式的设定办法也同 RNG 键。

⑩ 点数功能键（COU）

本天平设有点数功能，其平均数设有 5、10、25、50 四挡。

平均数范围设置：只要按 COU 键不松手，显示器就会如下所示，不断循环。

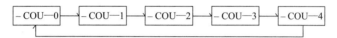

如需要一般称量功能，当显示器出现 COU—0 时即松手，随即出现等待状态————，最后出现称量状态 0.0000g。

如需要进入点数状态，当显示器出现 COU—1，COU—2，COU—3，COU—4 任意一种状态时即松手，显示器出现相应的显示状态，COU—05，COU—10，COU—25，COU—50 任意一种情况，分别代表为 5、10、25、50 只的平均值。

例如当显示器出现 COU—1 时松手，随即显示器显出 COU—05，其中"—05—"为闪烁状态，表示秤盘上放入 5 只被称物，再按一下 CAL 键，随即出————等待状态，约 8s 后显示器上出现 5，拿去被称物，显示器显示"0"，这时就可对被称物相同物体进行点数工作（注意：被称物体的质量不能大于天平的最大称量）。

若用 10、25，甚至 50 只进行平均，则点数的精度会更高些。

本天平由于具有断电记忆功能，所以若认为原有的平均数是准确的，则可免去平均功能操作步骤。操作如下：按住 COU 键，显示器出现 COU—1、COU—2、COU—3、COU—4 中任意一种状态即松手，按去皮键（TAR），随即显示器显示"0"，即可进行点数工作。

1.3.2.2　磁力加热搅拌器

磁力加热搅拌器可同时进行加热与搅拌，特别适合微型实验。搅拌的产生是通过磁场的不断旋转变化来带动容器中搅拌磁子的转动，转速可用调速器调节。搅拌时，调速器旋钮应慢慢旋转，过快会使磁子脱离磁场而不停地跳动，这时应立即将旋钮调到停位，待磁子停止跳动后，再逐步加速。

1.3.2.3　电热套与调压器

电热套是有机化学实验中常用的加热设备，调压器主要是用来控制电压，调节加热速度的设备。使用时要注意以下几点。（a）正确连接，保证它们的外壳不带电；（b）在使用过程中注意不能漏水,如果有水漏入,需及时停止加热;(c)在使用过程中电压由大到小逐步调节,停止加热时回到"0"。注意:不要长时间在大于或等于200V的电压下加热。

1.3.2.4 显微熔点仪和阿贝折光仪

二者将在第2章中进行详细阐述。

1.3.3 微型实验药品的称量

微型实验药品用量少,要求称量准确。对于液体药品需要量体积时,可使用移液管。固体或液体药品需要称重时,可直接在扭力天平或电子天平上称量。实验中如使用的溶剂量较大时,可用10mL量筒量取。所有的药品称量时,必须使用工具,不得用手直接触拿,必要时应戴防护手套。有些易挥发的有毒液体试剂,要在通风橱中取用。所有药品称取后均要及时加盖,以免药品吸潮、挥发、氧化而影响以后的使用。当用圆底烧瓶或茄形瓶称取药品时,可在天平盘上放一个玻璃圈来固定,以防烧瓶歪倒。

1.4 手册的查阅及有机化学文献简介

有机化学实验看起来涉及的主要内容是有机化学的知识,但实际上它所应用的知识涉及化学知识的方方面面,而且还要不断了解、掌握、运用和发现有关新知识。这样,在进行有机化学实验的时候,有必要经常查阅有关手册和文献,尤其是有机化学的文献。

有关有机化学的文献资料非常多,本书分工具书、期刊、化学文摘和网上资源4部分作简单介绍。如果需要了解详细的文献查阅知识,可参考有关专门的文献查阅书籍。

1.4.1 工具书和参考书

(1) Handbook of Chemistry and Physics

这是一本英文的化学与物理手册,于1913年出版第一版,2017年已经出到97版。内容分6个方面:数学用表;元素和无机化合物;有机化合物;普通化学;普通物理常数和其他。书中第三部分是有机化合物常数,有沸点、熔点、溶解度和相对密度等,按照有机化合物英文名称的字母顺序排列。在该部分的后面还有分子式索引,使用也很方便。

(2) The Merk Index

该书收集了一万余种化合物的性质、制法和用途。并提供了分子式索引和主题索引。

(3) Handbuch der Organischen Chemie (Beilstein)

该书一般称为贝尔斯坦有机化学大全,内容非常全面,提供了有关化合物的来源、制备方法、物理性质、化学性质、生理作用、用途和分析方法等,并有原始文献资料的出处,供进一步参考。共有31卷,分正编,第Ⅰ补编~第Ⅴ补编。该书有严格的编排原则,最简单的方法是根据其编排原则进行查找,也可根据分子式索引或主题索引进行查找。

(4) Vogel's Textbook of Practical Organic Chemistry, 5th Ed., 1996

这是一本较经典的有机化学实验教科书,对一些典型的基本操作、各类化合物的制备方法、定性分析等作了详细的介绍。

(5) 化工辞典(第五版)(化学工业出版社,2014年出版)

这是一本综合性的化工工具书,收集了有关化学、化工名词16000余条,列出了物质名词的分子式、结构式,基本的物理化学性质和有关数据,并有简要的制法和用途说明。

1.4.2 期刊

(1) 中文杂志

中文杂志与化学有关的非常多，主要有"中国科学"、"化学学报"、"化学通报"、"化学世界"、"高等学校化学学报"、"有机化学"等。

(2) 英文杂志

英文杂志与化学有关的也非常的多，重要的有：Journal of the American Chemical Society，Angewandte Chemie International Edition，Chemical Science，Organic Letters，The Journal of Organic Chemistry，Chemical Review，Chemical Society Reviews，Organic & Biomolecular Chemistry，Tetrahedron，Tetrahedron Letters 等。

1.4.3 化学文摘

化学文摘是将大量的、分散的各种文字的文献加以收集、摘录、分类整理而出的一种杂志。在众多的文摘性刊物中以美国化学文摘（Chemical Abstracts，简称 CA）最重要。CA 创刊于 1907 年，现在每年出两卷，每周一期。CA 的索引系统比较完善。有期索引、卷索引，每十卷有累积索引。累积索引主要有分子式索引（Formula Index）、化学物质索引（Chemical Substance Index）、普通主题索引（General Subject Index）、作者索引（Author Index）、专利索引（Patent Index）等。

1.4.4 网上资源

随着网络技术的迅速发展，网上的化学资源也日益丰富，使用也越来越方便。一般只要对网络知识有一定的了解，从网上查找有关化学资料是非常方便、迅速的。这里对有关资源作简单的介绍。

(1) 高校图书馆网站：例如 http：//lib.njtech.edu.cn；http：//lib.tsinghua.edu.cn；http：//www.lib.seu.edu.cn；http：//lib.nju.edu.cn 等。进入有关学校的图书馆网站可以查阅中国期刊网的有关资料，绝大多数中国期刊都上了"中国期刊网"，对有关期刊可从主题词、作者、期刊名称等方面查找。

(2) 中国国家图书馆：http：//www.nlc.cn。

(3) 中国知网：http：//www.cnki.net。由清华大学、清华同方发起，始建于 1999 年 6 月，是集期刊、博士论文、硕士论文、会议论文、报纸、工具书、年鉴、专利、标准、国学、海外文献资源为一体的、具有国际领先水平的网络出版平台。提供 CNKI 源数据库、外文类、工业类、农业类、医药卫生类、经济类和教育类多种数据库。其中综合性数据库为中国期刊全文数据库、中国博士学位论文数据库、中国优秀硕士学位论文全文数据库、中国重要报纸全文数据库和中国重要会议论文数据库。

(4) 万方数据知识服务平台：http：//wanfangdata.com.cn。可查基础科学、农业科学、人文科学、医药卫生和工业技术等众多领域的期刊；还有企业与产品、专业文献、期刊会议、学位论文、科技成果、中国专利等数据库。

(5) 专利检索及分析：http：//www.pss-system.gov.cn/。

(6) SciFinder Scholar：SciFinder Scholar 是美国化学学会所属的化学文摘服务社 CAS (Chemical Abstract Service) 出版的化学资料电子数据库学术版，它是全世界最大、最全面的化学和科学信息数据库。如前所述，《化学文摘》（CA）是涉及学科领域最广，收集文献类型最全，提供检索途径最多，部卷也最为庞大的一部著名的世界性检索工具。

CA 网络版 SciFinder Scholar，整合了 Medline 医学数据库、欧洲和美国等 30 几家专利机构的全文专利资料，以及化学文摘 1907 年至今的所有内容。涵盖的学科包括应用化学、化学工程、普通化学、物理、生物学、生命科学、医学、聚合体学、材料学、地质学、食品科学和农学等诸多领域。

SciFinder Scholar 收集由 CAS 出版的数据库的内容以及 MEDLINE 数据库，所有的记录都为英文（但如果 MEDLINE 没有英文标题的，则以出版的文字显示）。它有以下 5 个数据库：

Reference Database（文献数据库）：包含来自 150 多个国家、9500 多种期刊的文献，覆盖 1907 年到现在的所有文献以及部分 1907 年以前的文献，包括期刊、专利、会议录、论文、技术报告、书等，涵盖化学、生化、化学工程以及相关学科，还有尚未完全编目收录的最新文献。目前有大于 3000 万条参考书目记录，每天更新 3000 条以上。

Structure Database（结构数据库）：涵盖从 1957 年到现在的特定的化学物质，包括有机化合物、生物序列、配位化合物、聚合物、合金、片状无机物。REGISTRY 包括了在 CASM 中引用的物质以及特定的注册，例如：管制化学品列表如 TSCA 和 EINECS 中的注册。目前有大于 7400 万条物质记录，每天更新约 7 万条，每种化学物质有唯一对应的 CAS 注册号。

Reaction Database（反应数据库）：包括从 1907 年到现在的单步或多步反应信息。CASREACT 中的反应包括 CAS 编目的反应以及下列来源：ZIC/VINITI 数据库（1974～1991，by InfoChem GmbH），INPI（Institut National de la Propriete Insutrielle，法国）1986 年以前的数据，以及由教授 Klaus Kieslich 博士指导编辑的生物转化数据库。目前有大于 800 万条反应记录和大于 40 万条文献记录，每周更新约 700～1300 条。

Commercial Sources Database（商业来源数据库）：化学品的来源信息，包括化学品目录手册以及图书馆等内供应商的地址，价格等信息。目前有大于 740 万条商业化学物质记录，来自 655 家供应商的 793 种目录。

Regulatory Database（管制数据库）：1979 年到现在的管制化学品的信息，包括物质的特征、详细目录、来源以及许可信息等。有大于 22.8 万种化合物的详细清单，来自 13 个国家和国际性组织，每周更新 50 多条新记录。

SciFinderScholar 有多种先进的检索方式，比如化学结构式（其中的亚结构模组对研发工作极具帮助）和化学反应式检索等。它还可以通过 Chemport 链接到全文资料库以及进行引文链接。其强大的检索和服务功能，有助于了解最新的科研动态，确认最佳的资源投入和研究方向。根据统计，全球 95% 以上的科学家们对 SciFinder 给予了高度评价，认为它加快了他们的研究进程，并在使用过程中得到了很多启示和创意。

SciFinder Scholar 使用的简单介绍：

主要分为 Explore、Saved Searches 和 SciPlanner。如图 1-8 所示。

图 1-8 SciFinder Scholar 主界面

① Explore
② Explore Tool 可获取化学相关的所有信息及结构等，如图 1-9。
（a）Chemical Substance or Reaction-Retrieve the corresponding literature；
（b）By chemical structure；
（c）By substance identifier；
（d）By molecular formula；
（e）By Markush；
（f）By property；
（g）By reaction structure；
（h）Research Topic—to find literature relevant to a topic of interest；
（i）Author Name—to locate literature written by a specific author；
（j）Document Identifier—to find literature for a specific CA accession number or patent number；
（k）Company Name—to locate literature for a specific company，university，governmental agency，or other organization；
（l）Journal—to locate literature for a specific journal；
（m）Patent—to locate literature for a specific patent；
（n）Tags—to locate literature for tags which are user-defined keywords or terms that you can associate with a reference. You can manage tags from both a reference answer set and the reference detail. You can also indicate whether to export or print tags with your reference（s）。

图 1-9　Explore Tool 界面

③ Saved Searches，如图 1-10 所示。

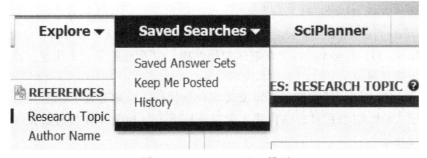

图 1-10　Saved Searches 界面

可在 Saved Searches 下 Saved Answer Sets 浏览文献的所有答案集，如果对感兴趣的文献做了警示标记，可在 Saved Searches 中的 Keep Me Posted 下找到，还可以在 History 中找到历史浏览记录。

④ SciPlanner 是一个由动态空间构建的合成方法、文献、物质和反应的搜索结果页面。如图 1-11 所示。

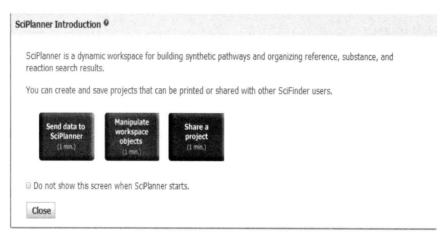

图 1-11　Sciplanner 搜索结果界面

(7) Reaxys 数据库

Reaxys 数据库由爱思唯尔（Elsevier）公司出品，是一个内容非常丰富的化学数值与事实数据库。2011 年起，Beilstein/Gmelin 数据库的访问从 Crossfire 平台转至 Reaxys 平台。Beilstein 和 Gmelin 为当今世界上最庞大和享有盛誉的化合物数值与事实数据库，编辑工作分别由德国 Beilstein Institute 和 Gmelin Institute 进行。前者收集有机化合物的资料，后者收集有机金属与无机化合物的资料。Reaxys 将贝尔斯坦（Beilstein）、专利化学数据库（Patent）和盖墨林（Gmelin）的内容整合为统一的资源，包含了 2800 多万个反应、1800 多万种物质、400 多万条文献。

它包含以下 3 个数据库。

① CrossFire Beilstein Database：世界最全的有机化学数值和事实数据库。时间跨度从 1771 年至今；包含化学结构相关的化学、物理等方面的性质；包含化学反应相关的各种数据；包含详细的药理学、环境病毒学、生态学等信息资源。

② Patent Chemistry Database：选自 1869~1980 年的有机化学专利；选自 1976 年以来有机化学、药物（医药、牙医、化妆品制备）、生物杀灭剂（农用化学品、消毒剂等）、染料等的英文专利（WO、US、EP）。

③ CrossFire Gmelin Database：全面的无机化学和金属有机化学数值和事实数据库。时间跨度从 1772 年至今；包含详细的理化性质，以及地质学、矿物学、冶金学、材料学等方面的信息资源。

Reaxys 数据库基于网络访问，无需安装客户端软件。检索界面简单易用，可以用化合物名称、分子式、CAS 登记号、结构式、化学反应等进行检索，并具有数据可视化、分析及合成设计等功能。

(8) Web of Science

Web of Science 由 Thomson 公司出版，是《科学引文索引（SCI）》的网络版，收录 5900 余种期刊文摘和引文，内容涉及自然科学、工程技术的各个领域。Web of Science 是一个基于 Web 所建立的整合的数字研究环境，为不同层次、不同学科领域的学术研究人员提供信息服务。目前，Web of Science 推出了增强的跨库联合检索功能，用户不仅可以同时检索所在机构订购的基于 Web of Science 平台上的所有资源，而且可以通过其提供的跨库联合检索功能，检索数据不在其平台上但对研究者来说非常重要的学术信息资源。目前该平台包括以下 8 个常用数据库。

① Web of Science—Science Citation Index Expanded　Web of Science 是 Thomson 建设的三大引文数据库的 Web 版，由三个独立的数据库组成（既可以分库检索，也可以多库联合检索），分别是 Science Citation Index Expanded（简称 SCIE）、Social Sciences Citation Index（简称 SSCI）和 Arts & Humanities Citation Index（简称 A&HCI）。内容涵盖自然科学、工程技术、社会科学、艺术与人文等诸多领域内的 8500 多种学术期刊。其中的 SCIE 数据库-《科学引文索引》网络版，收录 11000 余种期刊文摘和引文，内容涉及自然科学、工程技术的各个领域。数据每周更新。

② ISI Journal Citation Reports（Science Edition）　期刊引用报告（Journal Citation Reports 简称 JCR）依据来自 Web of Science（Science Citation Index Expanded 和 Social Sciences Citation Index）中的引文数据，提供可靠的统计分析方法，对全球学术期刊进行客观、系统地评估，帮助用户以定量的方式了解全球的学术期刊，并且通过这些分析数据可以了解某本学术期刊在相应研究领域中的影响力。JCR 从世界上经同行评议的学术期刊中，筛选出被引次数最高的近 8000 种期刊，涵盖了 200 多门学科，提供自 1997 年以来的期刊引文统计分析数据。用户可以根据需要，对检索到的期刊群进行以下各种方式的排序：影响因子、立即指数、总引用次数、刊载论文总数、被引半衰期或期刊名称。JCR 具有两个版本：自然科学版覆盖 ISI 数据库中的 6100 多种国际领先的科学技术领域期刊，社会科学版覆盖 ISI 数据库中约 1800 多种国际领先的社会科学领域期刊。中国科学院订购并开通了该数据库的自然科学版。

③ Current Contents Connect　Current Contents Connect 包括近 8000 种经同行评议的学术期刊，还有涉及自然科学、社会科学以及艺术人文学科的 100 多个学科的 1900 多本书籍和 5000 多个网站；可分为七个子库：生命科学（Life Sciences）1360 多种期刊；临床医学（Clinical Medicine）1200 多种期刊；工程学、计算与科技（Engineering, Computing & Technology）1150 多种期刊；农学、生物学与环境科学（Agriculture, Biology & Environmental Sciences）1100 多种期刊；物理、化学与地球科学（Physical, Chemical & Earth Sciences）1160 多种期刊；社会与行为科学（Social & Behavioral Sciences）1690 多种期刊；艺术与人文学科（Arts & Humanities）1100 多种期刊；和两个合集：商业合集（Business Collection）230 多种期刊，电子与电信合集（Electronics & Telecommunications Collection）200 多种期刊。

④ Biosis Previews　Biosis Previews（简称 BP）是著名的生物学信息检索工具之一，由美国 BIOSIS 公司出版。它由 Biological Abstracts（BA）、Biological Abstracts/ RRM（Reports, Reviews and Meeting）组合而成。数据库来自 5000 多种期刊以及国际会议、综述性文章、书籍、专利文献、图书和报告等，收录的文献涵盖世界 90 多个国家和地区。数据最早可回溯到 1969 年；记录总数超过 1450 万条。数据每周更新，每年新增数据量超过 56 万条。数据还包括来自美国专利商标局的 21000 条专利信息，这些专利的年代为 1986—1989 和 1995 至今。

⑤ Proceedings Proceedings通过网络方式提供会议论文的书目信息和作者摘要，其收集著名国际会议、座谈会、研讨会、讲习班和学术大会上发表的会议论文，覆盖了从1990年至今召开的60000次会议上发表的约200万篇论文，学科涉及自然科学和社会科学两大领域，覆盖250个学科，包括工程学、物理学、生物学、化学、精神病学和经济学等。数据库每周更新，每年新增超过385000条记录。

⑥ Derwent Innovations Index Derwent Innovations Index（简称DII）是由Thomson Scientific出版的基于Web的专利信息数据库。Derwent Innovations Index 4.0将Derwent World Patents Index和Patents Citption Index有机地整合在一起，用户不仅可以通过它检索专利信息，而且可以通过这个数据库检索到专利的引用情况，还可以利用Derwent Chemistry Resources展开化学结构检索。该数据库收录了来自全球40个专利发行机构的1200多万个基本发明，专利覆盖范围可追溯到1963年，引用信息可追溯到1973年，是检索全球专利最权威的数据库。

⑦ Highly cited.com 该数据库依据ISI数据库中各学科领域的总被引次数收录科研人员信息，主要是来自21个科学领域的核心人物，包括：神经系统科学、工程学、物理、化学、计算机科学、地球科学、分子生物学、遗传学和空间科学等。目前每个学科领域的入选者是250个。网站每周更新，提供科学家的个人信息、研究兴趣和出版物信息。该资源为免费资源。

⑧ Medline Medline是美国国立医学图书馆（NLM）的书目信息数据库，收录1950年至今的生物医学和生命科学、生物工程学、公共健康、临床护理以及植物与动物科学方面的生命科学文献。拥有来自生命科学领域30种语言、4900多个全球出版物中期刊文章的近1300万个参考书目，此资源包括了期刊、报纸以及时事通讯。通过ISI Web of Knowledge平台，用户可以使用或不使用MesH词表以及NLM可控词表叙词检索Medline；也可以使用标准化学名称或CAS登记号检索化学物质。此外，它们还能连接到NCBI数据库获取基因序列信息，并可以从Medline全记录链接到PubMed相关文章。

Web of Science的使用界面如下，主要分为检索、检索分析和保存，如图1-12所示。

图1-12 Web of Science使用界面

（a）检索 检索类型包括基本检索，被引参考文献检索和高级检索。检索项目包括：主题，标题，作者，作者识别号，编者，团体作者，出版社名称和DOI号等。检索时间可以通过时间跨度来设置。

(b) 检索分析　以"graphene"为关键字做主题检索，结果如图 1-13 所示。

图 1-13　Web of Science 检索分析

左上角显示相关文献数目，下面是对检索结果限定的工具栏。

结果限定选项包括：出版年，研究领域，数据库，文献类型，基金资助机构，作者等。

另外可以对检索结果进行排序：出版日期（降序），出版日期（升序），最近添加，被引频次（降序），被引频次（升序），使用次数-最近 180 天，使用次数-2013 年至今。

(c) 保存　如图 1-14 所示，点击文献序号前的小方块，然后点击标记结果列表就可以保存相应的文献。

图 1-14　Web of Science 检索结果的保存

网上的文献资源数不胜数，这里不一一列举。只要多上网，根据以上网址就很容易查找到相关网址与资料。注意：SciFinder Scholar 和 Web of Science 需要所在单位购买相关使用许可权才可以使用。

2 基本操作与注意事项

2.1 蒸馏和分馏

蒸馏是提纯和分离有机化合物的常用方法。如果将液体加热，随着温度的升高，蒸气压将增大，当液体蒸气压与外界大气压相等时，液体内气泡逸出即为沸腾，此时的温度称为该液体的沸点。通常所说的沸点就是指在1个大气压（101.325kPa，即760mmHg）下液体沸腾的温度。在这种情况下的蒸馏操作称为常压蒸馏，简称蒸馏。

蒸馏是将液体加热到沸腾，使之汽化，然后将蒸气冷凝为液体的过程。通过蒸馏，可使混合物中沸点不同的各组分得到部分或全部分离。分离效果取决于各组分之间的沸点差，沸点差越大，分离效果越好。一般来讲，各组分之间沸点差在30℃以上时，才能获得较好的分离效果。

然而，对于沸点相差不大的混合物，难以用普通蒸馏的方法将它们分离。如果采用分馏柱，当混合蒸气进入分馏柱时，高沸点组分首先被冷凝，冷凝液中就会含有较多的高沸点组分。而低沸点组分在蒸气中就相对增多。冷凝液向下流动时，与上升的蒸气接触，二者进行热量交换，结果又使上升蒸气中高沸点组分被冷凝，低沸点组分仍呈蒸气上升。而在冷凝液中，低沸点组分则受热不断气化，高沸点组分仍呈液态下流，如此多次的液相与气相的热量交换，使得低沸点组分不断气化上升，最终被蒸馏出来，高沸点组分则不断流回容器中，从而使沸点相差不大的混合物得以有效地分离、纯化，这就是分馏。实际上，分馏就是在分馏柱中进行多次的部分气化、冷凝过程。但是，对于恒沸混合物，无论分馏效率有多高，也不能用分馏的方法把它们分离开，除非用其他方法破坏共沸组分后，才能用分馏方法来纯化。

2.1.1 蒸馏和分馏装置

首先根据电热套位置调节十字夹高度，夹好圆底烧瓶，烧瓶不要太低，应以方便地将电热套取出、放入为宜。为此可在电热套下放置垫板，使圆底烧瓶在电热套中距内壁1～2cm，严禁烧瓶触及电热套。液体加入量为圆底烧瓶体积的1/3～2/3，烧瓶口接H形分馏头，其正上口装温度计，调节温度计的高度使水银球上端位置与分馏头横侧管下限位置处在同一水平线上。竖侧管向上的开口接直形（或球形）冷凝管，冷凝管下支管为进水口，上支管为出水口。H形分馏头竖侧管末端接锥形瓶或圆底烧瓶作为接收瓶，见图2-1。

蒸馏前，烧瓶中应加沸石，以防止液体暴沸[1]。如忘加沸石，必须待液体冷却后再补加。间断蒸馏，每次都要加沸石。开始加热前，要先通冷凝水，并检查各部分仪器连接是否紧密，严防漏气。

2.1.2 蒸馏和分馏操作

在圆底烧瓶中，加入蒸馏物，其量控制在烧瓶体积的1/3～2/3之间，然后加1～2粒沸石。按图 2-1 安装好仪器，开通冷凝水[2]，调节电压开始加热。刚开始时，电压可稍高一些，待液体接近沸腾时，调低电压，维持沸腾，使液体馏出速度为每秒 1～2 滴，在整个蒸馏过程中应该使温度计水银球上常有冷凝的液体，此时的温度就是液体（馏出物）的沸点，蒸馏速度太快或太慢，温度计的读数均不准。

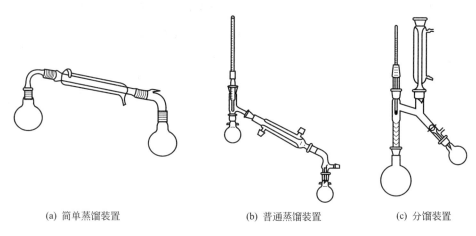

(a) 简单蒸馏装置　　　　　　(b) 普通蒸馏装置　　　　　　(c) 分馏装置

图 2-1　蒸馏和分馏装置

蒸馏过程中，注意观察温度的变化，待温度趋于稳定（或达到所需温度）时，关闭 H 形分馏头活塞，更换接收瓶（在此之前的馏出物为"前馏分"）。蒸馏完毕，先把调压器指针调回零，然后拔下电源插头，撤掉电热套，停止通冷凝水，拆卸仪器（其程序与安装程序相反），最后将所有仪器清洗干净。

【附注】

[1] 液体的温度超过自身的沸点仍不沸腾时，这种现象称为"过热"。此时，一旦有一个气泡形成，由于液体在此温度下蒸气压已远远超过大气压和液柱压力之和，因此上升的气泡增大特别快，甚至会将液体溢出瓶外，这种不正常的沸腾现象称为"暴沸"。

[2] 冷凝水的流速以能使蒸气充分冷凝为宜，通常只需保持缓慢的水流即可，对于高沸点（大于 140℃）的物质，蒸馏时不要通冷凝水。

思 考 题

(1) 什么叫分馏？在什么情况下采用分馏操作？蒸馏和分馏有什么不同？
(2) 什么叫沸点？沸点与气压的关系是什么？温度计的位置如何调节才能准确显示馏分沸点？
(3) 如果一个液体具有恒定的沸点，是否能断定它就是单纯物质？
(4) 蒸馏过程中应注意哪些问题？
(5) 沸石的作用是什么？忘记加沸石时，若不将液体冷却就加入会发生什么现象？
(6) 蒸馏时，为什么要控制馏出液的速度为每秒 1～2 滴？

2.2　减压蒸馏

减压蒸馏是分离提纯那些在常压蒸馏时沸点较高，或在常压蒸馏时未达沸点即分解、氧化或聚合的物质的常用方法。

液体的沸点是指其蒸气压与外压相等时的温度。因此液体的沸腾温度与外界压力有关。

外界压力增大，沸腾温度升高；外界压力减小，沸腾温度降低。利用此原理可以在较低的压力下对某些物质进行蒸馏，避免样品发生分解、氧化和聚合等副反应。这种在较低压力下进行的蒸馏操作，称为减压蒸馏。

表 2-1 列出了几种化合物在不同压力下的沸点，可以看出，在压力降低到 2.666kPa（20mmHg）时，大多数有机化合物的沸点比常压下的沸点要降低 100～120℃左右。

表 2-1　几种化合物的压力与沸点的关系

压力 /kPa(mmHg)	沸 点 /℃					
	水	氯苯	苯甲醛	水杨酸乙酯	甘油	蒽
101.325(760)	100	132	179	234	290	354
6.665(50)	38	54	95	139	204	225
3.999(30)	30	43	84	127	192	207
3.325(25)	26	39	79	124	188	201
2.666(20)	22	34.5	75	119	182	194
1.995(15)	17.5	29	69	113	175	186
1.333(10)	11	22	62	105	167	175
0.665(5)	1	10	50	95	156	159

当减压蒸馏在 1.333～3.325kPa（10～25mmHg）之间进行时，压力相差 0.133kPa（1mmHg），沸点相差大约 1℃。

进行减压蒸馏时，首先要知道某一压力下化合物的沸点，下述 3 种途径可以得到有关数据：

（1）查阅文献。

（2）查压力-沸点关系曲线。图 2-2 所示为液体在常压下沸点与减压下沸点的近似关系图。若已知某物质在 101.325kPa（760mmHg）下的沸点值，由此经验曲线便可找出在减压至某值时该物质的沸点。如二乙基丙二酸二乙酯常压下沸点为 218～220℃，欲减压至 2.666kPa（20mmHg），可在中间线和右边线上分别找到上述两点，连成一直线并延长，使其与左边线相交，交点便是此压力下液体的沸点，约为 105～110℃。

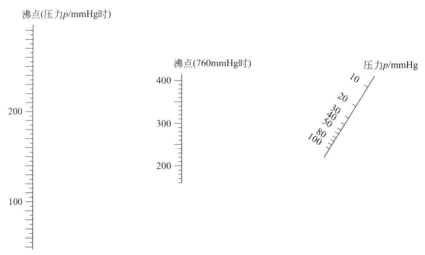

图 2-2　液体在常压下的沸点与减压下的沸点的近似关系

1mmHg＝0.133kPa

(3) 公式计算。已知两组压力与温度值，可由下式求得常数 A 和 B。代入式中便可求出任一压力下的相应沸点。

$$\lg p = A + \frac{B}{T}$$

式中，p 为蒸气压；T 为热力学温度（$T = 273.2 + t℃$），K。

2.2.1 减压蒸馏装置

2.2.1.1 减压装置

实验室中，减压装置常作为一个整体装置出现，它还可以分为抽气、保护、测压 3 个部分。

① 抽气部分　实验室中常用水泵或油泵进行减压，尤其是油泵。水泵所能达到的最低压力为该温度下水的蒸气压。油泵的效能取决于油泵的机械结构以及油的好坏（即油的蒸气压的高低）。蒸馏时挥发性的有机溶剂蒸气、水和酸的蒸气等都会损坏油泵，因此在使用时必须十分注意保护油泵。

② 保护装置　当用油泵进行减压时，为防止易挥发的有机溶剂、酸性物质和水蒸气进入油泵，必须在馏出液接收器和油泵之间顺次安装冷阱和几种吸收塔，以免污染油泵用油，腐蚀机件。冷阱置于盛有冷却剂的广口保温瓶中。冷却剂的选择视需要而定，可用冰-水、冰-盐、干冰等。吸收塔通常设两个，前一个装无水氯化钙或硅胶，后一个装粒状氢氧化钠。有时为吸除有机溶剂，可再加一个石蜡片或活性炭吸收塔。

③ 测压部分　通常采用水银压力计[1]和数字式压力计[2]来测量减压系统的压力。

2.2.1.2 蒸馏装置

减压蒸馏的蒸馏装置需要实验者自己安装。它由二口圆底烧瓶、H 形分馏头、温度计、减压毛细管、直形（或球形）冷凝管、温度计套管、空心塞及接收瓶组成。其安装与常压蒸馏装置类似。二口圆底烧瓶正口接 H 形分馏头，侧口装减压毛细管。毛细管的作用是使液体均匀地沸腾。它的顶端接一带有螺旋夹的橡皮管，管中插一细铜丝，毛细管的尖端插入液面下，距烧瓶底 1～2mm 为宜。减压毛细管通过橡皮管、温度计套管与二口圆底烧瓶连接。接收瓶选用圆底烧瓶或茄形瓶，不能用平底烧瓶或锥形瓶。冷凝管上口加盖空心塞。减压系统连在 H 形分馏头活塞下部向上的支管上。

在蒸馏部分和减压部分之间有一安全瓶，起防护作用，瓶口装有二通活塞。整个仪器装置如图 2-3 所示。实验中，根据馏分沸点高低，可分别选用（a）、（b）装置。

实验开始时，先进行减压，检测系统内压力，若达不到所需真空度（仪器本身问题除外），应检查是否有漏气处，然后停止减压，在仪器连接处涂抹真空脂。真空脂用量不要太多，一般涂在标准磨口上半部，形成均匀薄层，再将两仪器连接，旋转几下，磨口变得透明即可。

2.2.2 减压蒸馏操作

(1) 样品加到二口圆底烧瓶中（体积在 1/3～1/2），如图 2-3(a) 安装好仪器。

(2) 旋紧减压毛细管上的螺旋夹，打开安全瓶的二通活塞，开泵抽气。然后再关闭二通活塞，从水银压力计上观察系统所能达到的真空度。如因漏气而不能达到所需真空度，应停泵处理后再抽气减压。

(3) 估计收集温度范围，调节螺旋夹，使液体中有连续平稳的小气泡通过（如无气泡，

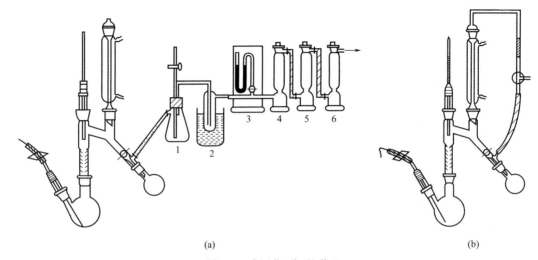

图 2-3 减压蒸(分)馏装置
1—安全瓶；2—冷阱；3—压力计；4—氯化钙塔；5—氢氧化钠塔；6—石蜡塔

要更换减压毛细管，更换时停止抽气）。开通冷凝水，用电热套加热蒸馏。调节适宜的电压，控制液体馏出速度为每秒1~2滴。整个蒸馏过程中，都要密切注意温度计和压力计的示数。收集沸点范围一般不超过预期温度±1℃。

（4）当温度计读数发生变化或液体几乎蒸干时，停止蒸馏。此时，应先移去热源，稍冷后，渐渐打开安全瓶的二通活塞，使系统与大气相通，同时松开毛细管上的螺旋夹（防止液体吸入毛细管），当系统内外压力平衡后，停止抽气，关闭冷凝水，卸下接收瓶。

若减压过程中发现有前馏分，应在接近预期温度时，撤掉电热套，关闭H形分馏头活塞，打开安全瓶活塞，更换接收瓶。此后，再关闭安全瓶活塞，打开H形分馏头活塞，继续加热，收集馏分，此时应注意动作迅速。另外，也可缓缓打开毛细管顶端螺旋夹和安全瓶活塞，使其与大气相通，待体系内外压力平衡后，再更换接收瓶。

【附注】

[1] 水银压力计

水银压力计有封闭式和开口式两种，如图2-4和图2-5所示。

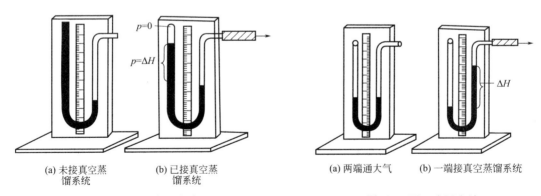

(a) 未接真空蒸馏系统　(b) 已接真空蒸馏系统

图 2-4 封闭式压力计

(a) 两端通大气　(b) 一端接真空蒸馏系统

图 2-5 开口式压力计

实验室中常用开口式压力计，它测量的是大气压[3]与系统压力之差。因此，蒸馏系统内的实际压力（真空度）应是大气压减去汞柱高度，即

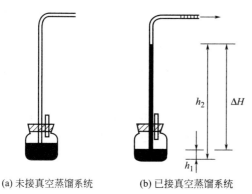

(a) 未接真空蒸馏系统　　(b) 已接真空蒸馏系统

图 2-6　简易开口式压力计

$p_{系统} = p_{大气压} - \Delta H_{压力计}$

有时为了方便，实验中通常采用简易开口式压力计，见图 2-6。汞柱高度 $\Delta H = h_2 - h_1$，减压蒸馏系统内的真空度计算方法与用开口式压力计完全相同。

[2] 数字式压力计

数字式压力计前面板如图 2-7 所示。

① 电源键：共有关机、AC、DC 三种工作位置，每按键 3 次，依次循环变换 1 次。

(a) 关机位置：电源切断，指示灯和数码管均不亮。

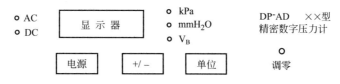

图 2-7　数字式压力计前面板示意图

(b) AC 位置：红色 AC 指示灯亮。已接 220V 市电，数码管同时亮；反之，只有 AC 指示灯亮，数码管不亮。

(c) DC 位置：绿色 DC 指示灯和数码管同时亮，此时由机内电池供电。

② +/- 键：每按 1 次，显示器上的极性符号变化 1 次。测正压和差压时，使读数为正；测负压时，使读数为负。

③ 选择键：共有 kPa、mmH_2O 和 V_B 三种工作位置，每按键 3 次，依次循环变换 1 次。

(a) kPa 位置：kPa 指示灯亮，示值为以 kPa 为单位的压力值。

(b) mmH_2O 位置：mmH_2O 指示灯亮，示值为以 mmH_2O 为单位的压力值。

(c) V_B 位置：V_B 指示灯亮，示值为以伏为单位的机内电池电压值。

④ 调零器：测压前，用以调零。

数字式压力计后面板如图 2-8 所示。

图 2-8　数字式压力计后面板示意图

(a) 压力接嘴：量程在 600kPa 以上者只有 1 个压力接嘴；250kPa 以下者，有正压和负压接嘴各 1 个，分别用以引入正压力和负压力。若测差压，高压端接"+"，低压端接"-"。

(b) BCD 码输出：DP-ADD 型有 BCD 码和控制信号输出，共 18 线。从左到右为：Com、B_1、B_2、B_4、B_8、D_4、D_3、D_2、D_1、Busy、CLK、PoL、D_5、R/H、STR、OR 和 UR。

(c) 24V 输出：DP-ADI 型的压力示值为 4~20mA，后面板上备有 24V 电源，可引出供被测的变送器用。

操作步骤：

① 将本仪表的压力接嘴接至被测/被校系统。600kPa 以下的仪表用内径为 ϕ4~5 的塑料管连接，600kPa 以上的仪表，用专用高压管连接。

② 接上电源。当用 220V 市电时，用随机所附的三芯电源线将本机接入市电，当用直流工作时，则不接电源。按电源键，将电源接通。交流供电时，红色 AC 指示灯亮；直流供电时，绿色 AC 指示灯亮。

③ 根据需要，按选择键选择压力单位 kPa 或 mmH_2O。相应的指示灯亮。

④ 预热 10~15min，调零，使读数为 "0000"。

⑤ 气密性检查。加压或疏空至量程上限值，关闭压力源，观察 1min，检查被测系统气密性。若气密性良好，则读数不变或只有几个字变化；若读数的绝对值不断下降，则有漏气。只有在确定系统气密性良好后，才能继续进行测试。

⑥ 检查气密性后，卸压回零。此时读数为 "0000"，若有偏离，重新调零。

⑦ 示值检测。经上述步骤后，即可进行正式检测校验工作。其方法步骤与用活塞压力计相似，不再赘述。

⑧ 工作完毕，关机。

(a) 因工作需要用 220V 供电，可以不关机长期连续工作；但直流供电，必须关机，以免电池消耗。

(b) 未接市电，按键关机，误至 "AC" 位置；或用市电工作，拔了 220V 插头而忘了关机，仍处于 "AC" 位置，此时数码管虽灭，但 "AC" 指示灯仍亮。这将消耗机内电池。这种状态超过 1 周将使机内电池耗尽而损坏，必须防止。关机后，必须检查 "AC" 指示灯是否熄灭。

⑨ 充电。在市电供电工作时，同时对机内电池充电；充足电后，自动断电停充。

⑩ 机内电池状况检查。机内电池电压充足电后约为 6.3~6.4V，其放电终止电压约为 5.2~5.4V。到终止电压后，自断停放，数码管熄灭。此时，必须换到 "AC" 位置进行充电。

到野外工作之前，应检查电池消电状态。方法是按选择键 V_B 位置，由所显示的电压值进行判别。电压在 6V 以下，应将电池充足。

[3] 大气压力计

精密空盒气压表如图 2-9 所示，由该气压表的指针可直接读取大气压力。其使用方法是：仪器工作时必须水平放置，防止任意方向的倾斜而造成仪器读数的误差；为了消除传动机构中的摩擦，在读数前轻敲仪器的外壳或玻璃；读数时观察者的视线必须与度盘平面垂直；气压和温度的读数均必须精确到小数点后 1 位；对气压读数需进行刻度、温度的补充校正，方法如下。

温度校正：

$$\Delta p_1 = \Delta p \times t$$

式中，Δp 为温度变化 1℃时的气压校正值（仪器的检定证

图 2-9 精密空盒气压表

书上有）；t 为温度计读数，$p = p' + \Delta p_1 + \Delta p_2 + \Delta p_3$；$p'$ 为读数值；Δp_2 为刻度校正值（仪器的检定证书上有）；Δp_3 为补充校正值（仪器的检定证书上有）。

思 考 题

（1）哪些化合物必须采用减压蒸馏方法纯化？为什么？真空度高低对化合物的沸点有什么影响？
（2）减压蒸馏时为什么用长毛细管代替沸石？为什么不能用锥形瓶或平底烧瓶作接收瓶？
（3）怎样正确掌握减压蒸馏操作顺序？为什么先抽气减压，待系统压力恒定后再开始加热蒸馏？为什么不先加热再减压？
（4）停止蒸馏时，为什么要先移去热源，再使系统与大气相通，待系统内外压力平衡后，才停止抽气？
（5）试画出整个减压蒸馏装置图。

2.3 水蒸气蒸馏

水蒸气蒸馏是提纯有机混合物的重要方法之一。当水和不（或难）溶于水的化合物共存时，根据道尔顿分压定律，整个体系的蒸气压为各组分蒸气压之和，即 $p = p_A + p_B$。式中 p 为总的蒸气压，p_A 为水的蒸气压，p_B 为不溶于水的化合物的蒸气压。当 p 等于外界大气压时，混合物开始沸腾，所以混合物的沸点将低于其中任一组分的沸点。这样，在常压下进行水蒸气蒸馏，就可在 100℃ 以下将高沸点组分与水一同蒸出。馏出液经冷却、分离，便可得到纯度较高的物质。

混合物蒸气压中各气体分压之比等于它们的物质的量之比，即

$$\frac{p_A}{p_B} = \frac{n_A}{n_B}$$

又

$$n_A = \frac{m_A}{M_A}, \quad n_B = \frac{m_B}{M_B}$$

所以

$$\frac{n_A M_A}{n_B M_B} = \frac{p_A M_A}{p_B M_B} = \frac{m_A}{m_B} \quad [1]$$

式中，m_A、m_B、M_A、M_B 分别代表物质 A、B 蒸气的质量和相对分子质量。

由上式可见，两种物质在馏出液中的相对质量与它们的蒸气压及相对分子质量成正比。水有较低的相对分子质量和较高的蒸气压，其乘积较小，这样就可以用来分离较高相对分子质量和较低蒸气压的物质，甚至固体也可用水蒸气蒸馏来纯化。

水蒸气蒸馏适用于以下几种情况：
① 含大量树脂状杂质或不挥发物，用蒸馏、萃取等方法难以分离的混合物。
② 常压下普通蒸馏会发生分解的高沸点有机物。
③ 脱附混合物中被固体吸附的液体有机物。
④ 除去易挥发的有机物。

运用水蒸气蒸馏时，被提纯物质应具备以下条件：
① 不溶或难溶于水。
② 沸腾温度下不与水反应。
③ 在 100℃ 左右，必须具有一定的蒸气压（一般不少于 1.333kPa 即 10mmHg）。

实验装置如图 2-10(a) 所示[2]。在 1 只二口圆底烧瓶中加入样品萘 3.0g，再加 12mL 水和一粒沸石，按图 2-10(a) 安装好仪器并开始加热，接收瓶用冷水浴冷却。冷凝水不宜开

得太大,以防萘蒸气凝固堵塞冷凝管。若发现内壁有白色固状物出现时,关闭冷凝水[3],让热蒸气把它加热熔化。若仍然堵塞,则停止加热,取下冷凝管,用玻璃棒将它捅出,然后重新安装,加入沸石,继续加热蒸馏,待馏出液不再浑浊而转为澄清[4]时,停止加热,取下接收瓶,冷却至室温,抽滤。产品在红外灯下烤干或晾干。称重,测熔点,计算回收率。

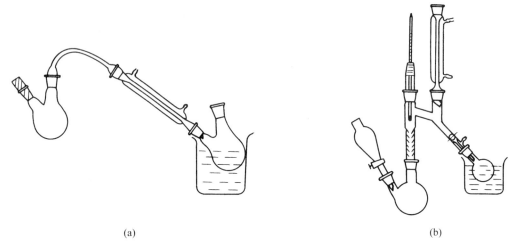

图 2-10 水蒸气蒸馏装置

【附注】

[1] 此公式只适用于不溶于水的物质,而实际上很多化合物在水中都或多或少有一定的溶解。因此,这样的计算只是近似的。

[2] 本实验是利用水蒸气蒸馏提纯固体。若被提纯物质为液体,则图 2-10(a)(b) 均可选用。

[3] 为了防止过多的萘凝固堵塞冷凝管,可提前关闭冷凝水,需要时再开通。

[4] 如果馏出液仍浑浊,可在蒸馏瓶中添加水,继续蒸馏,直至馏出液变清。

思 考 题

(1) 在什么情况下,必须用水蒸气蒸馏方法来提纯物质?举例说明。

(2) 水蒸气蒸馏时,如何判断有机物已完全馏出?

(3) 今有硝基苯、苯胺的混合物,能否利用水蒸气蒸馏的方法将二者分离?为什么?

(4) 以下几组混合体系中,哪几个可用水蒸气蒸馏法(或结合化学方法)进行分离?

a. $CH_3C_6H_4Br$ 与 $CH_3C_6H_4NH_2$

b. $CH_3CH_2CH_2OH$ 与 $CH_3—CH—CH_3$
                                              $|$
                                             OH

c. Fe,$FeCl_3$,C_6H_5Cl

2.4 熔 点 测 定

所谓物质的熔点是指该物质的液相和固相之间处于平衡状态时的温度。纯物质的熔点为固定值,它是物质的重要物理性质之一,可以用来鉴定固体物质的纯度。如果将杂质加入纯的有机化合物中,将产生熔点下降、熔程增长的现象。因此,利用混合熔点法可以鉴别混合物中各组分是否相同。如果熔点不变，则为相同物；如果熔点下降，熔程增长，则为不同物。但生成固熔体或新化合物的例外也是有的。

实际上测定的熔点不是单一的温度，而是固体熔化的温度范围，即物质的熔程，指记录固体最初开始熔融的温度到全部熔融的温度。

熔点的测定有毛细管法和显微熔点测定仪法，这里分别介绍如下。

2.4.1 毛细管法测定熔点

(1) 样品的装入

将少许干燥样品放在干净的表面皿上用玻璃棒或不锈钢刮刀将它研成粉末。将市售毛细管在酒精灯上熔融封闭一端，开口端向下插入粉末中，制成熔点管，然后把熔点管开口端向上，轻轻地在桌面上敲击，以使粉末落入和填紧管底。最好取1支长30～40cm的玻璃管垂直于干净的表面皿上，将熔点管从玻璃管上端自由落下，可更好地达到上述目的。如此重复数次，以使管内装入高约2～3mm紧密结实的样品。粘于管外的粉末须拭去，以免玷污加热浴液。要测得准确的熔点，样品一定要研得极细，装得结实，使热量的传导迅速均匀。对于蜡状的样品，为了解决研细及装管的困难，可选用较大口径（2mm左右）的毛细熔点管。

(2) 熔点浴

熔点浴的设计最重要的目的是要使受热均匀，便于控制和观察温度。下面介绍两种在实验室中最常用的熔点浴。

① 提勒管式（Thiele，又称b形管） 如图2-11(a)所示。管口装有开口软木塞，温度计插入其中，刻度应面向木塞开口，其水银球位于b形管上下两叉管口之间，装好样品的熔点管借少许浴液沾附于温度计下端，使样品的部分置于水银球侧面中部[见图2-11(c)]，再剪取一小段橡皮管，将此橡皮圈套在温度计和熔点管的上部[见图2-11(d)]，使样品管不掉入加热液体（浴液）中。b形管中装入浴液，高度达上叉管处即可。在图示的部位加热，受热的浴液沿管上升运动，从而在整个b形管内浴液呈对流循环，使样品受热较为均匀。

② 双浴式 如图2-11(b)所示。将试管经开口软木塞插入250mL平底（或圆底）烧瓶内，直至离瓶底约1cm处，试管口也配一个开口橡皮塞或软木塞，插入温度计，其水银球

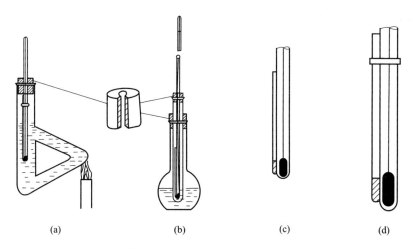

图 2-11　毛细管法测定熔点装置
(a) 提勒管熔点装置；(b) 双浴式熔点装置；(c)、(d) 毛细管的位置与安装方法

应距试管底 0.5cm。瓶内装入约占烧瓶 2/3 体积的加热液体，试管内也放入一些加热液体，使插入温度计后，其液面高度与瓶内相同。熔点管沾附于温度计水银球旁，与在 b 形管中相同。

在测定熔点时，凡是样品熔点在 220℃ 以下时，可采用浓硫酸作为浴液。长期不用的熔点浴应先慢慢加热去掉吸入的水分。若加热过快，硫酸浴液有冲出的危险。当有机物和其他有机杂质混入硫酸浴液时，会使硫酸变黑，有碍熔点的观察，此时可加入少许硝酸钾晶体共热后使之脱色。

除浓硫酸以外，亦可采用磷酸（可用于 300℃ 以下）、石蜡油或有机硅油等。如将 7 份浓硫酸和 3 份硫酸钾或 5.5 份浓硫酸和 4.5 份硫酸钾在通风橱中一起加热，直至固体溶解，这样的溶液可应用在 220~320℃ 的范围。若以 6 份浓硫酸和 4 份硫酸钾混合，则可使用至 365℃。但此类加热液体不适用于测定低熔点的化合物，因为它们在室温下呈半固态或固态。

（3）熔点的测定

将装好样品的提勒管垂直夹于铁架上，以浓硫酸作为加热液体，以小火在图示部位缓缓加热。开始时升温速度可以较快，到距离熔点 10~15℃ 时，调整火焰使每分钟上升约为 1~2℃。愈接近熔点，升温速度应愈慢（**注意：升温速度是准确测定熔点的关键**）。一方面是为了保证有充分的时间让热量由管外传至管内，以使固体熔化；另一方面因观察者不能同时观察温度计度数和样品的变化情况，只有缓慢加热，才能使此项误差减小。记下样品开始塌落并有液相（俗称出汗）产生时（初熔）和固体完全消失时（全熔）的温度计读数，即为该化合物的熔程。要注意，在初熔前必须仔细观察是否有萎缩或软化、放出气体以及其他分解现象。例如某物质在 120℃ 时开始萎缩，在 121℃ 时有液滴出现，在 122℃ 时全部液化，应记录如下：熔点为 121~122℃，120℃ 时萎缩。

熔点测定，至少要有 2 次重复的数据。每一次测定都必须用新的熔点管另装样品，不能将已测过熔点的熔点管冷却，使其中的样品固化后再做第二次测定。因为有时某些物质会产生部分分解，有些会转变成具有不同熔点的其他结晶形式。测定易升华物质的熔点时，应将熔点管的开口端烧熔封闭，以免升华。如果要测定未知物的熔点，应先对样品粗测 1 次。加热速度可以稍快，知道大致的熔点范围后，待浴温冷至熔点以下约为 30℃，再取另一根装样的熔点管作精密的测定。熔点测好后，温度计的读数须对照温度计校正图进行校正。

2.4.2 显微熔点测定仪测定熔点

图 2-12 是 X-4 显微熔点测定仪[1]的示意图。操作步骤如下。

（1）对新购置的仪器。最好先用熔点标准药品进行测量标定。求出修正值（修正值＝标准药品的熔点标准值－该药品的熔点测定值），作为测量时的修正依据。

（2）对待测物品进行干燥处理。把待测物品研细，放在干燥器内，用干燥剂干燥；或者用烘箱直接快速烘干（温度应控制在待测物品的熔点温度以下）。

（3）将熔点热台放置在显微镜底座 ϕ100 孔上，并使放入盖玻片的端口位于右侧，以便于取放盖玻片及药品。

（4）将熔点热台的电源线接入调压测温仪后侧的输出端，并将传感器插入熔点热台孔，其另一端与调压测温仪后侧的插座相连；将调压测温仪的电源线与 220V 电源相连。

（5）取两片盖玻片，用蘸有乙醚（或乙醚与酒精混合液）的脱脂棉擦拭干净。晾干后，取适量待测物品（不大于 0.1mg）放在 1 片载玻片上并使药品分布薄而均匀，盖上另 1 片载玻片，轻轻压实，然后放置在熔点热台中心。

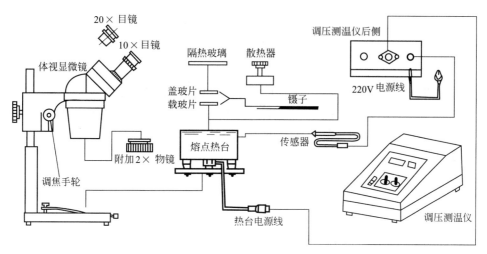

图 2-12 X-4 显微熔点测定仪的示意图

(6) 盖上隔热玻璃。

(7) 松开显微镜的升降手轮,参考显微镜的工作距离(88mm 或 33mm),上下调整显微镜,直到从目镜中能看到熔点热台中央的待测物品轮廓时锁紧该手轮;然后调节调焦手轮,直至能清晰地看到待测物品的像为止[2]。

(8) 打开电源开关,调压测温仪显示出熔点热台即时的温度值(**注意:测试操作过程中,熔点热台属高温部件,一定要使用镊子夹持放入或取出。严禁用手触摸,以免烫伤!**)。

(9) 根据被测物体熔点的温度值,控制调温手钮 1 或 2(1 表示升温电压宽量调整,2 表示升温电压窄量调整,其电压变化可参考电压表的显示),以期达到在测量物质熔点过程中,前段升温迅速、中段升温渐慢、后段升温平缓。具体方法如下:先将两个调温手钮顺时针调到较大位置,使熔点热台快速升温。当温度接近待测物体熔点温度以下 40℃ 左右时(中段),将调温手钮逆时针调节至适当位置,使升温速度减慢。在被测物熔点值以下 10℃ 左右(后段)时,调整调温手钮控制升温速度约每分钟 1℃(**注意:尤其是后段升温的控制对测量精度影响较大。当温度上升到距待测物熔点值以下 10℃ 左右,经过反复调整手钮 1 或 2,一定要将升温速度控制在大约每分钟 1℃。方便的无级调整会让用户很快掌握,运用自如**)。

(10) 观察被测物品的熔化过程,记录初熔和全熔时的温度值[3],用镊子取下隔热玻璃和盖玻片,即完成 1 次测试。如需重复测试,只需将散热器放在熔点热台上,电压调为零或切断电源,使温度降至熔点值以下 40℃ 即可。

(11) 对已知熔点的物质,可根据所测物质的熔点值及测温过程,适当调节调温旋钮,实现测量;对未知熔点物质,可先用中、较高电压快速粗测 1 次,找到所测物质熔点的大约值,再根据读数适当调整和精细控制测量过程,最后实现较精确的测量[4]。

(12) 精密测试时,对实测值进行修正,并多次测试,计算平均值。

物品熔点值的计算如下。

一次测试:

$$T = X + A$$

式中,T 为被测物品熔点值;X 为测量值;A 为修正值。

多次测试：

$$T = \frac{\sum_{i=1}^{n}(X_i + A)}{n}$$

式中，T 为被测物品熔点值；X_i 为第 i 次熔点值；A 为修正值；n 为测量次数。

(13) 测试完毕，应及时切断电源，待熔点热台冷却后，方可将仪器按规定装入包装。

(14) 用过的载玻片可用乙醚擦拭干净，以备下次使用。

【附注】

[1] 仪器应置于阴凉、干燥、无尘的地方使用与存放。

[2] 透镜表面有污秽时，可用脱脂棉蘸少许乙醚和乙醇混合液轻轻擦拭，遇有灰尘，可用洗耳球（吹球）吹去。

[3] 不要把熔程记录为初熔与全熔温度读数之差。

[4] 易升华和易氧化样品的测定应用两端封闭的毛细管，将毛细管直接放在加热板上进行测定。

思 考 题

(1) 如何用测定熔点的方法来判断化合物的纯度？

(2) 今有固体化合物 A、B，且两者熔点相同，能否断定 A、B 为同一物质？如何断定？

2.4.3 Melting Point Determination of Benzoic Acid

【Experimental Objectives】

(1) Learn the principle and method of melting point determination.

(2) Master the determination of melting range of benzoic acid crystals.

【Experimental Principles】

The melting point (m. p.)[1] of a substance is the temperature in which the balance between the liquid and solid phase. The melting point of a pure substance is a fixed value, and it can be used to determine the purity of a solid substance. In this experiment, the melting point of benzoic acid was measured by a micro melting point apparatus, and the initial melting temperature and total melting temperature of benzoic acid should be noted. The temperature gap between total melting and initial melting are named the melting range.

【Apparatus and Reagents】

(1) Apparatus: Micro melting point apparatus[2], cover glass, absorbent cotton, tweezers.

(2) Reagents: Benzoic acid, diethyl ether.

【Experimental Procedure】

Take two pieces of cover glasses, clean them with cotton dipped with ether[3]. After drying, take some amount of benzoic acid (no more than 0.1mg) and disperse them on one piece of cover glass uniformly. Then cover with another one, press it down gently, and place it at the center of the melting point plat of the apparatus.

Adjust the microscope's lift hand wheel until you can clearly see the object. Switch on the power (220V) and adjust the temperature.

The process of initial melting and total melting of benzoic acid was observed and their temperature values should be recorded[4].

Repeat the above steps three times.

【Notes】

[1] Actually, the term "melting point" is a misnomer. You must always report a melting range, from the temperature at which melting starts to the temperature at which it finishes. Melting points should be reported to the nearest whole degree. The melting point of a pure compound should be over a narrow range (1—3℃). An impure compound will have a broader range. In addition, impurities tend to lower the melting point of a substance.

[2] The equipment should be kept in a cool, dry and dust-free place for use and storage.

[3] When the lens surface has dirt, it can be gently wiped with the absorbent cotton dipped in a little ether and ethanol. In case of dust, blow it away with the rubber suction bulb.

[4] Do not record the melting range as the difference between initial and full melting temperature readings.

【Thinking Questions】

(1) How can the purity of a compound be determined by measuring the melting point?

(2) There are solid compounds A and B with same melting points, can we determine A and B the same substances? How to determine?

(3) Think about the Arctic Ocean. How does the water up there stay liquid at below 0℃?

2.5 重 结 晶

重结晶是纯化固体有机物的重要方法之一，它是利用被提纯化合物及杂质在所选择的溶剂中，在不同温度下溶解度的不同，通过热过滤除去不溶性固体物质、活性炭吸附有色杂质及冷却后舍去含有杂质的母液3种途径，除去杂质，从而达到纯化目的。

2.5.1 重结晶过程

（1）将不纯的固体有机化合物在溶剂的沸点或接近沸点的温度下溶解在溶剂中，制成接近饱和的浓溶液。若固体有机化合物的熔点较溶剂沸点低，则应制成在熔点温度以下的近饱和溶液。

（2）若溶液含有有色杂质，可加活性炭煮沸脱色。

（3）将上述溶液进行热过滤，以除去其中的不溶物及活性炭。

（4）将滤液冷却，使结晶从过饱和溶液中析出，而可溶性杂质留在母液中。

（5）抽滤。从母液中将结晶分出，洗涤晶体以除去被吸附的母液。

（6）干燥。晶体干燥后测定熔点，以检测其纯度。如不符合要求，可重复上述操作，直至检验合格。

2.5.2 溶剂的选择及用量

进行重结晶时，选择合适的溶剂是一个关键问题。理想的溶剂应具备以下条件：

① 不与被提纯物质起化学反应。

② 在较高温度下能溶解多量被提纯物质，而在室温或更低温度下只能溶解少量。

③ 对杂质的溶解度非常大（留在母液中除去）或非常小（热过滤除去）。
④ 容易挥发，易与结晶分离除去。
⑤ 能给出较好的结晶。

重结晶常用溶剂见表 2-2。

表 2-2 重结晶常用溶剂

溶 剂	沸点/℃	冰点/℃	相对密度	与水的互溶性	易燃性
水	100	0	1.0	+	0
甲醇	64.96	<0	0.7914	+	+
95%乙醇	78.1	<0	0.804	+	++
冰醋酸	117.9	16.7	1.05	+	+
丙酮	56.2	<0	0.79	+	+++
乙醚	34.5	<0	0.71	-	++++
石油醚	30~60	<0	0.64	-	++++
乙酸乙酯	77.06	<0	0.90	-	++
苯	80.1	5	0.88	-	++++
氯仿	61.7	<0	1.48	-	0
四氯化碳	76.54	<0	1.59	-	0
粗汽油	65~75	<0	0.7	-	+++
环己烷	81	6.5	0.8	-	+++

注："+"表示与水的互溶性好或易燃性好；"-"表示与水的互溶性差。

具体选择溶剂时，一般化合物可先查阅手册中溶解度一栏。如无文献资料可查，只能用实验方法确定。具体做法是，把少量（约 0.1g）样品放入试管中，慢慢滴入溶剂，振摇，当加入溶剂量约为 1mL 时，小心加热并摇动，观察加热和冷却时样品的溶解情况（加热时严防溶剂着火）。若化合物在 1mL 冷或温热的溶剂中已完全溶解，则此溶剂不适用。如该化合物不溶于 1mL 沸腾的溶剂，可继续加热，再滴入溶剂，如超过 4mL 仍不溶解，则此溶剂也不适用。如果该化合物能溶解在 1~4mL 沸腾的溶剂中，可将试管冷却，观察结晶析出情况。若结晶不能自行析出，可用玻璃棒摩擦溶液液面以下的试管壁，或置于冰水中冷却，使结晶析出。若结晶仍不能析出，则此溶剂不适用。如果结晶能正常析出，要注意结晶量的大小，可以用几种溶剂同法进行比较，选用结晶收率最高、操作容易、毒性小、价格便宜的溶剂来进行重结晶。如果在 0.1g 固体中，加入 0.5~1mL 溶剂，在加热时能将样品完全溶解，冷却时又能以 80%~90%左右的回收率获得良好的晶体时，一般即可认为该溶剂相宜。

当一种物质在一些溶剂中溶解度太大（这种溶剂被称为良溶剂），而在另一些溶剂中溶解度又太小（这种溶剂被称为不良溶剂），不能选择到一种合适的溶剂时，用混合溶剂往往可得到满意的结果。所谓混合溶剂，就是把对该物质溶解度很大和溶解度很小且又互溶的两种溶剂混合起来。

用混合溶剂重结晶时，可先将待纯化的物质放在接近沸腾的良溶剂中溶解。若有不溶物，趁热过滤；若有色，用活性炭煮沸脱色并趁热过滤。在此热溶液中小心地加入热的不良溶剂，直至出现混浊并不再消失，再加入少量良溶剂或稍加热使其恰好透明，然后经冷却、结晶、抽滤，便得到纯化的物质。有时也可将两种溶剂先行混合，再按单一溶剂进行操作。

常用的混合溶剂如下：

乙醇-水、乙酸-水、丙酮-水、吡啶-水、乙醚-甲醇、乙醚-丙酮、乙醚-石油醚、苯-石油醚。

溶剂加入量应比需要量多20%左右[1]。

2.5.3 溶解及热过滤

溶解时，为避免溶剂挥发及可燃性溶剂着火或有毒溶剂逸出，应在烧瓶上安装回流冷凝管，装置见图2-13。先加较需要量略少的溶剂，加热沸腾。若未完全溶解，可分次逐渐添加溶剂，再加热到沸腾并摇动，直到刚好溶解为止。但要注意判断是否有不溶或难溶的杂质存在，以免误加过多的溶剂。若含有有色杂质，可加活性炭沸腾脱色。活性炭的脱色能力只在极性溶剂（特别是含羟基的溶剂）中才能充分发挥，在烃类溶剂（如石油酸）中效果甚差。加活性炭时应注意以下几点。

① 加活性炭前，首先将待结晶化合物溶液冷却到沸点以下。在接近沸点的溶液中加入活性炭易引起暴沸。

② 加入活性炭的量视杂质的多少而定，一般为粗品质量的1%～5%。过量的活性炭将吸附一部分产品。

③ 如一次脱色不完全，可重复进行。

④ 加入活性炭后，要煮沸5～10min，活性炭脱色后，趁热过滤，装置见图2-14[2]。

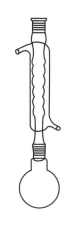

图2-13 回流装置

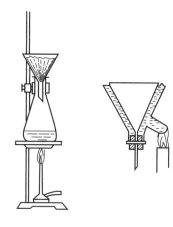

图2-14 热过滤装置

热过滤是重结晶成败的关键，应注意以下几点。

① 漏斗颈应短而粗，使用前在80℃左右的烘箱中预热。

② 漏斗中放一张折叠滤纸，折叠滤纸向下凸起的棱边贴紧在漏斗壁上，滤纸上沿不可超过漏斗口。用少量热溶剂湿润滤纸[3]，以免干滤纸吸收溶液中的溶剂，使结晶析出，堵塞滤纸孔。将热溶液倒在滤纸的上部，这样溶液与较大面积的折叠滤纸接触，过滤就比较快。但要特别留意不能让溶液从漏斗与滤纸之间漏过。在此过程中，为减少溶剂挥发，漏斗上盖一个表面皿，凹面向下。折叠滤纸的方法：将手用肥皂洗净，擦干。首先将滤纸对半折叠，再将其依次对半同向折叠成8等份，见图2-15(a)～(c)。然后在1和10，10和5，5和7，…，9和3间各反向折叠，如图2-15(d)。把滤纸打开，在1和3的地方各内折叠一个小叠面，最后做成如图2-15(e)的折叠滤纸。在折叠滤纸时，切记不要用手指甲刮折痕，特别是滤纸的中央部分，如折痕太紧，会减少其牢度以致在过滤时破裂。

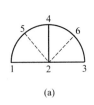 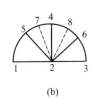 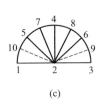 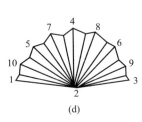 

图 2-15 折叠滤纸的方法

2.5.4 结晶、抽滤和干燥

2.5.4.1 结晶

滤液自然冷却，可慢慢析出颗粒较大的结晶。如无结晶析出，可用玻璃棒摩擦器壁或投入晶种（若无晶种，可用玻璃棒蘸取一些溶液，稍干后即会析出晶体）。有时从滤液分出的是油状物而不是结晶，油状物中常常含有不少杂质，因此要避免油状物的形成。对油状物的处理可采用两种方法。其一，刮擦瓶壁或投入少许原不纯固体的细小晶种于混合液中静置，如仍无结晶析出，分出油状物，改用其他溶剂重结晶。其二，有些油状物当温度再降低时，就固化成结实的硬块。遇此情况，可添加适量溶剂，将滤液重新加热至沸再冷却。

2.5.4.2 抽滤

在三角滤板漏斗上铺一块比滤板稍小的圆形滤纸，用少量的溶剂湿润，然后打开水泵，关上安全瓶二通活塞，抽吸滤纸使其紧贴于滤板上面。将容器中的液体和晶体借助玻璃棒分批倒入漏斗中，最后用母液冲洗器壁上的晶体，全部转移到漏斗中去，压紧抽干。

2.5.4.3 洗涤

打开安全瓶上的活塞，通入大气。用少量重结晶用的溶剂[4]润湿所有的晶体，并用玻璃棒轻轻搅动一下，再关上安全瓶的活塞，抽滤。把溶剂尽量抽净。一般重复洗涤 2 次即可。最后用玻璃钉将晶体压紧，抽干溶剂，用小刮刀将晶体和滤纸一并刮下，放在表面皿上，把晶体摊开。抽滤后的母液，若有必要，可回收或将其蒸发浓缩，便可得到一部分纯度较低的晶体。

2.5.4.4 干燥

通常用空气晾干、红外灯或烘箱烘干、滤纸吸干、在干燥器中干燥等方法。干燥后，将滤纸弃去，晶体放在干燥器中保存，待测纯度。

【仪器与药品】

(1) 仪器：20mL 圆底烧瓶，直形（或球形）冷凝管，短颈漏斗，三角滤板漏斗，玻璃钉，20mL 锥形瓶，10mL 锥形瓶，不锈钢刮刀，表面皿，20mL 烧杯。

(2) 药品：粗萘（工业品）1.0g，苯甲酸（粗品）0.5g，70% 乙醇 15mL，50% 乙醇 7mL，活性炭少量。

【步骤】

(1) 70% 乙醇重结晶粗萘

在装有回流冷凝管的 20mL 圆底烧瓶中，放入 1.0g 粗萘[5]，加入 7mL 70% 乙醇及 1 粒沸石，开通冷凝水，加热至沸腾[6]，并不时摇动，加速溶解。若萘不能完全溶解，可从冷凝管上口添加少量70%乙醇，并注意观察溶解情况，待完全溶解后，再多加3～4mL 70%乙醇。移去电热套，稍冷后加入少许活性炭并稍加摇动，重新加热回流5～10min。趁热用预热的短颈漏斗和折叠滤纸过滤。滤纸先用70%乙醇润湿，滤液用20mL锥形瓶接收（**注意：附近不要有明火！**）。滤完后用少量热的70%乙醇洗涤滤纸和烧瓶。锥形瓶加盖空心塞，令其自然冷却，最后用冷水冷却使结晶完全析出，用三角滤板漏斗抽滤。结晶用少量70%乙醇洗涤1次，用玻璃钉压紧，抽干溶剂，将结晶置于表面皿上，在空气中或红外灯下干燥。测产品熔点、称重、计算回收率。纯萘为白色片状晶体，熔点为80.55℃。本实验可得纯萘0.55～0.73g，回收率为55%～73%。

（2）用水重结晶苯甲酸

称取0.5g苯甲酸[7]，放入25mL圆底烧瓶中，加入10mL水和1～2粒沸石，用电热套加热至沸，并用玻璃棒不断搅动使固体完全溶解。如有未溶解的固体，可继续加入少量水，直至全部溶解。移去热源，稍冷后加入活性炭少许，继续加热煮沸5～10min。用预热好的短颈漏斗和折叠滤纸趁热过滤，滤纸用少量热水润湿。用20mL烧杯接收滤液。待所有溶液过滤完毕，用少量热水洗净烧瓶，倒入滤纸。滤液冷却至室温后，再用冷水冷却析出的结晶。用三角滤板漏斗抽滤晶体，并用玻璃钉挤压，使母液尽量滤去。用少量水洗涤晶体1～2次，重新压紧抽干。将结晶置于表面皿上，铺成薄层，在红外灯下烤干。测产品熔点、称重、计算回收率。

纯苯甲酸为无色针状晶体，熔点为122.4℃。本实验可得纯苯甲酸0.3～0.4g，回收率为60%～80%。

【附注】

[1] 过量的溶剂是为了防止热过滤转移沸腾的饱和溶液时因温度下降而引起结晶析出。

[2] 热过滤也可用抽滤的方法。

[3] 实验中也可采用在溶液中多加些溶剂的方法。

[4] 若重结晶溶剂沸点较高，一般用原溶剂洗涤1次后，再用低沸点溶剂（与原溶剂互溶但对晶体不溶或微溶）洗涤1次。

[5] 萘在乙醇中的溶解度见表2-3。

表 2-3　萘在70%乙醇中的溶解度

$t/℃$	10	20	25	30	40	50	60	70
g/100mL	0.7	0.93	1.04	1.13	1.52	2.70	4.44	8.33

[6] 萘的熔点较70%乙醇的沸点低，若加入不足量的70%乙醇，加热至沸后，萘呈熔融状态，并非溶解，这时应再加入溶剂直至完全溶解。

[7] 苯甲酸在水中的溶解度见表2-4。

表 2-4　苯甲酸在水中的溶解度

$t/℃$	0	10	20	30	40	60	70	80	90	95
g/100mL	0.17	0.21	0.29	0.42	0.60	1.2	1.77	2.75	4.55	6.8

思　考　题

（1）重结晶加热溶解时，为什么先加入比计算量略少的溶剂，然后再逐渐加至恰好溶解，最后再多加少量溶剂？

(2) 为什么活性炭要在固体物质全部溶解后再加入？活性炭脱色在什么溶剂中发挥效力最大？

(3) 以有机溶剂和以水为溶剂进行重结晶时，其在仪器装置和操作上各有什么不同？

(4) 如何选择溶剂？在什么情况下用混合溶剂？制备混合溶剂的先决条件是什么？

(5) 试考虑对空气、水敏感的物质，其重结晶操作应如何进行？

2.5.5　Recrystallization of Benzoic Acid

【Experimental Objectives】

(1) Learn and be familiar with the basic operation of solid dissolution, thermal filtration, decompression filtration and so on.

(2) The principle and significance of recrystallization purification of solid organic compounds were studied by benzoic acid recrystallization experiment.

【Experimental Principles】

A compound formed in a chemical reaction or extracted from some natural source is rarely pure when initially isolated. Organic chemists devote considerable effort to the isolation of pure products. Recrystallization of solids is a valuable technique to master because it is one of the methods used most often for purification of solids.

In solution recrystallization, advantage is taken of the fact that nearly all solids are more soluble in a hot solvent than in a cold solvent. If the crystals are dissolved in a quantity of hot solvent which is insufficient to dissolve them when cold, and if that hot solution is then allowed to cool, it should be anticipated that crystals will precipitate from the cooling solution to the extent of the difference in solubility between the temperature extremes. If the impurities present in the original crystals have dissolved and remain dissolved after the solution is cooled, filtration of the crystals which have formed on cooling should then provide purified material. Or if the impurities remain undissolved in the hot solution and are filtered from it before it is allowed to cool, the crystals which subsequently form on cooling should be more pure than the original crystals.

【Apparatus and Reagents】

(1) Apparatus: Beaker, round-bottom flask, heating mantle, Buchner funnel, suction flask, filter paper, weighing paper, watch glass, glass rod, measuring cylinder, circulating water pump, electronic balance.

(2) Reagents: Benzoic acid, zeolite, activated charcoal, water.

【Experimental Apparatus Graphs】

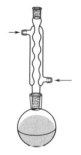 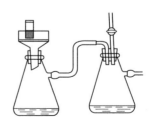

【Experimental Procedure】

Dissolution Place 1.0g of impure benzoic acid[1] in a 100mL flask, add 30mL of water, bring the mixture to the boiling point by heating it on a heating mantle. A sign indicating the water level is made on the flask wall (some hot water should be added to keep the water level for evaporation of water). Continue heating the mixture until the benzoic acid has dissolved completely.

Decoloration Pure benzoic acid is colorless, so a colored solution indicates that treatment with decolorizing activated charcoal is necessary[2]. (Caution: Do not add decolorizing activated charcoal to a boiling solution!) Cool the solution slightly, add a microspatula-tip full of activated charcoal, and reheat to boiling for a few minutes. Meanwhile prepare a funnel and a clean flask to receive the filtrate.

Hot filtration and crystallization Heat the funnel by inverting it in a water bath, clean with a clean towel, place the filter paper in the funnel, arrange the receiving flask to collect the hot filtrate. Without allowing the funnel or the solution to cool, pour the solution into the funnel. If the solution cannot be poured into the funnel in a single portion, replace it on the heating mantle and continue to heat it to prevent cooling. As soon as all of the solution has been filtered, cover the opening of the flask with a piece of filter paper, an inverted beaker, or loose-fitting cork to exclude airborne impurities from the solution, and allow the filtrate to stand undisturbed until it has cooled to room temperature and no more crystals form. To complete the crystallization, the flask can be placed in an ice-water bath for at least 15min.

Isolation and drying Collect the crystals on a clean Buchner or Hirsch funnel by vacuum filtration and wash the filter cake with two small portions of cold water. Press the crystals as dry as possible on the funnel with a clean cork or spatula. Spread the benzoic acid on a watch glass, protecting it from airborne contaminants with a piece of filter paper, and air-dry it at room temperature or in an oven. Be certain that the temperature of the oven is below the melting point of the product!

Analysis Determine the melting points of the crude and recrystallized benzoic acid, the weight of the latter material, and calculate the percent recovery using following equation.

$$\text{Percent recovery} = \frac{\text{weight of pure crystals recovered}}{\text{weight of original sample}} \times 100\%$$

The procedure of recrystallization are exemplified in Fig. 2-16.

【Notes】

[1] During the portion of this experiment when you are pouring or transferring solutions, either wear rubber gloves or be particularly careful to avoid getting these solutions on your skin. Organic compounds are much more rapidly absorbed through the skin when they are in solution, particularly in water-soluble solvents such as ethanol, acetone, and others. It is for this reason also that you should never rinse organic materials off your skin with solvents such as acetone instead, wash your hands thoroughly with hot water and soap.

[2] Do not add decolorizing activated charcoal to a boiling solution because it may cause

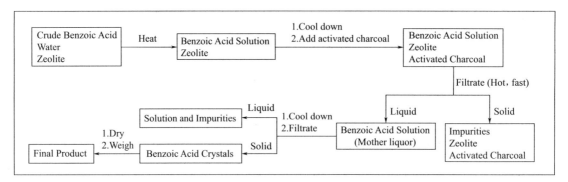

Fig. 2-16　Procedure of recrystallization

the mixture to boil out of the container.

【Thinking Questions】

(1) List the steps in the systematic procedure for miniscale recrystallization, briefly explaining the purpose of each step.

(2) The goal of the recrystallization is to obtain purified material with a maximized recovery. For each of the items listed, explain why this goal would be adversely affected.

1) In the solution step, an unnecessarily large volume of solvent is used.

2) The crystals obtained after filtration are not washed with fresh cold solvent before drying.

3) The crystals referred to in 2) are washed with fresh hot solvent.

4) A large quantity of decolorizing activated charcoal is used.

5) Crystals are obtained by breaking up the solidified mass of an oil that originally separated from the hot solution.

6) Crystallization is accelerated by immediately placing the flask of hot solution in an ice-water bath.

(3) Why decolorizing activated charcoal should not be added to a solvent that is at or near its boiling point?

(4) A second crop of crystals may be obtained by concentrating the filtrate and cooling. Why is this crop of crystals probably less pure than the first crop?

(5) How is the purity of a recrystallized solid assessed?

2.6　折射率的测定

光线在密度较大的透明液体中的传播速度比在空气中小，当光线从空气进入稠密介质时，传播速度减小，这就造成了光线的折射，如图 2-17 所示。

光线在空气中的速度（u）与它在液体中的速度（u'）之比定义为该液体的折射率 n：

$$n = \frac{u}{u'}$$

实验证明：光线在空气中的速度与在液体中的速度之比等于入射角 ρ 的正弦（$\sin\rho$）与折射角 θ 的正弦（$\sin\theta$）之比，因此，可以通过下式来计算液体的折射率：

$$n = \frac{\sin\rho}{\sin\theta}$$

液体的折射率与液体的温度和所用的光线有关，通常用 n_D^t 表示，D 表示用钠光灯的 D 线（589.3nm）作光源，t 是与折射率相对应的温度。温度升高，折射率降低，一般用一个近似公式[1]来计算。为了方便，通常采用 4×10^{-4}（$3.5\times10^{-4}\sim 4.5\times10^{-4}$）为温度变化常数，即温度升高 1℃，液体有机化合物的折射率应降低 4×10^{-4}。在实际工作中，往往把某一温度下测得的折射率换算成另一温度下的折射率。

测定折射率最常用的仪器是阿贝（Abbe）折光仪[2]。它的优点是：仪器用日光照射，通过光学补偿（消色散装置）得到相当于钠光源 D 线时的折射率，其测得数值与钠光源一样。阿贝折光仪的结构见图 2-18。

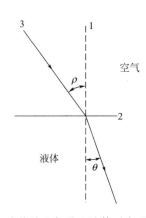
图 2-17　光线从空气进入液体时向垂线偏折图
1—垂线；2—液体界面；3—入射光线；
ρ—入射角；θ—折射角

图 2-18　阿贝折光仪的结构

对于纯净液体，折射率可精确到万分之一。通常用四位有效数字记录，它是纯净液体鉴定方法中较为精确的一种。

将阿贝折光仪与恒温槽相连接，恒温后，分开锁紧扳手，用擦镜纸蘸少量无水乙醇或丙酮轻轻擦洗上下镜面。待乙醇或丙酮挥发后，加 1 滴蒸馏水于下面镜面上，关闭扳手，调节反光镜，从目镜中观察，使镜内视场变亮，转动手轮直到棱镜内观察到有界线或彩色光带出现，若出现彩色光带，则调节手轮，使明暗界线清晰，再转动手轮，使界线恰好通过"十"字交叉点，记录读数与温度。重复 2 次得纯水的平均折射率，并与纯水标准值对照，可求得折光仪的校正值。然后以同样的方法测定乙醚和乙酸乙酯的折射率，校正步骤在实际操作中可省略。

纯水标准值：n_D^{20} 为 1.33299；纯乙醚标准值：n_D^{20} 为 1.3526；纯乙酸乙酯标准值：n_D^{20} 为 1.3723。

【附注】

[1] 近似公式为：
$$n_D^{20}=n_D^t+0.00045\times(t-20℃)$$
即把在 t℃时测得的折射率校正到在 20℃时的折射率。

[2] 使用阿贝折光仪应注意以下几点：

a. 阿贝折光仪量程从 1.3000～1.7000，精密度为±0.0001。测量时应注意温度是否正确，如欲精确至±0.0001，则温度应控制在±0.1℃范围内。

b. 折光仪在使用和贮藏时,均不应暴露于日光中,不用时用黑布罩住。

c. 折光仪的棱镜必须注意保护,不能在镜面上刻痕;滴加液体时,滴管末端切不可触及棱镜面。

d. 在每次滴加样品前,应洗净镜面。使用完毕,要用丙酮或无水乙醇清洗镜面,待晾干后,再关上棱镜。

e. 对棱镜玻璃、保温套金属以及其间胶合剂有腐蚀或溶解作用的液体,均应避免使用。

思 考 题

(1) 液体的折射率高低与哪些因素有关?

(2) 使用阿贝折光仪应注意哪些事项?

2.7 萃 取

萃取是利用化合物在两种互不相溶或微溶的溶剂中溶解度不同或分配系数不同,使某种化合物从一种溶剂中分配到另一种溶剂中一部分,经过反复多次萃取,使绝大部分化合物被提取出来。

分配定律是萃取理论的主要依据。在一定温度下,某化合物与两种溶剂不发生分解、电离、缔合和溶剂化等作用时,此化合物在两液层中含量之比是一定值,用公式表示为:

$$K = \frac{c_A}{c_B}$$

式中,K 为分配系数;c_A、c_B 表示一种化合物在两种互不相溶的溶剂中的物质的量浓度。

在有机化学实验中,经常遇到的是利用有机溶剂提取溶解在水中的有机化合物。但用水来提取有机化合物中不要的组分也是常有的事。例如,有机层中含有需要除去的酸,就可用水或稀碱水溶液来提取;如果有机层中含有要除去的碱性物质,就要用稀酸水溶液提取,通常,把这种过程称为洗涤。为了减少有机物在水中部分溶解的损失,萃取时多用饱和氯化钠水溶液代替水,或在被提取水溶液中加入一定量的电解质(如氯化钠),利用"盐析效应"来降低有机物或有机萃取剂在水中的溶解度,以提高萃取效率。

利用分配定律的关系,可以计算出经过萃取后化合物的剩余量。

设 V 为原溶液体积;m_0 为萃取前化合物的总量;m_1 为萃取 1 次后化合物的剩余量;m_2 为萃取 2 次后化合物的剩余量;m_n 为萃取 n 次后化合物的剩余量;S 为萃取剂的体积。

经过 1 次萃取,原溶液中化合物的浓度为 m_1/V;而萃取剂中化合物的浓度为 $(m_0-m_1)/S$;两者之比等于 K,即

$$\frac{m_1/V}{(m_0-m_1)/S} = K$$

整理得

$$m_1 = m_0 \frac{KV}{KV+S}$$

同理,经过 2 次萃取后,则有

$$m_2 = m_0 \left(\frac{KV}{KV+S}\right)^2$$

经过 n 次萃取后，有

$$m_n = m_0 \left(\frac{KV}{KV+S}\right)^n$$

由上述结果不难得出，同样体积的萃取剂，采取少量多次的方法，萃取效率要比一次萃取高，效果好得多。

从溶液中萃取物质的方法有两种，一种是分次萃取法，另一种是连续萃取法。

2.7.1 分次萃取法

应用分液漏斗进行分次萃取是实验室中常用的方法之一。一般选择比提取液体积大 1~2 倍的分液漏斗，把活塞擦干并涂上一层薄薄的润滑脂，塞好后，再把活塞沿着一个方向旋转数圈，使润滑脂均匀分布。将被提取溶液和萃取剂（一般为溶液体积的 1/3）依次自上口倒入分液漏斗中，塞紧塞子。右手掌顶住漏斗上口的玻璃塞，手指握住漏斗颈部，左手握着漏斗的活塞部分，大拇指和食指按住活塞柄，中指垫在塞座下边，将漏斗稍倾，使漏斗活塞部分向上（见图 2-19），慢慢振摇。每摇几次，在保持漏斗向上倾斜的情况下，慢慢打开活塞进行放气，待漏斗中过量气体逸出后，将活塞关闭，继续振摇。重复数次后，将漏斗置于台架上静置，待两层液体完全分开后，打开上口塞子，再将活塞慢慢旋开，进行分液。下层液体自分液漏斗下口放出，上层液体自上口倒出。多次提取中，再将被萃取层倒回分液漏斗中，重复操作。为使分液完全，当上层液体流近活塞时，先关闭活塞，停止分液，从台架上取下漏斗，用手握住活塞下部，轻轻按同一方向旋摇几下，再放在台架上静置分液。

图 2-19 分液漏斗的振摇

萃取时，如遇到乳化现象，两相相对密度相似难以分离时，可采取以下方法处理。

(1) 长时间静置。
(2) 加入少量电解质破乳或增大水相的相对密度，使两相分离。
(3) 若因溶液呈碱性而产生乳化，可加入少量稀硫酸或采用过滤方法除去。
(4) 根据不同情况，可以加入其他破乳物质，如乙醇、磺化蓖麻油等。

2.7.2 连续萃取法

如果被提取的物质在母液中的溶解度极大，用溶剂分次萃取效果极差时，为减少溶剂用量，可采用连续萃取方法。有关仪器装置可按图 1-1 的固液提取器安装。

思 考 题

(1) 用分液漏斗萃取时，为什么要放气？
(2) 用分液漏斗分离两相液体时，应如何分离？为什么？
(3) 用乙醚萃取水溶液中的有机物，有时在水层与醚层之间有许多泡沫，泡沫应分到水层还是留在醚层？在分液过程中，有泡沫沾到漏斗内壁，如何处理才能使有机相与水相分离完全？

(4) 若用有机溶剂萃取水溶液，而又不能确定有机溶剂在分液漏斗的哪一层，应如何做出迅速决定？

(5) 实验室中常用乙醚、氯仿、四氯化碳、苯作萃取剂萃取水溶液，说明它们在上层还是下层？

2.8 干　燥

实验室制备的药品以及工业生产产生的化学品往往因为反应、转移、储存等原因含有水分，水分的存在有两种方式：游离水分、结合水分。干燥过程最重要的影响来自于湿组分的性质及其与待干物质的结合方式；湿组分为水时，待干物质与水的结合方式决定其干燥方式。

2.8.1 液体的干燥

有机液体的干燥按照干燥强度（不同含水量要求）可以分为干燥和无水处理（即深度干燥）。

2.8.1.1 有机溶液的干燥

有机实验中，样品的处理往往包含分液→干燥→浓缩的操作。经过分液后的有机溶液含水量较大，向其中加入具有吸水性的固体药品（如 $CaCl_2$、$MgSO_4$、Na_2SO_4 等）进行吸水，此类盐型化合物就是通常所说的干燥剂，它们通过吸水转变成相应的结晶水合物。常见干燥剂的使用特点见表 2-5。

表 2-5　常见干燥剂的使用特点

干燥剂	水合物形式	使用温度	特点及适用对象
$CaCl_2$	$CaCl_2 \cdot 6H_2O$	<30℃	干燥作用较慢。适用对象：烃类、卤代烃。不适用于醇、酚、胺、羧酸等有机物溶液
$MgSO_4$	$MgSO_4 \cdot 7H_2O$	<48℃	干燥效率高、效果佳。适用对象：烃类、酯、醛、酮、酰胺、腈等。不适用于醇、酚、胺、羧酸等有机物溶液
Na_2SO_4	$Na_2SO_4 \cdot 10H_2O$	<30℃	吸水量大，但干燥不彻底，一般用于含水量较大的体系。几乎对所有液体都适用
K_2CO_3	$K_2CO_3 \cdot 2H_2O$		干燥效率、吸水量中等，一般适用腈、酮、酯、醇等，不适于含酸、酚组分的干燥

选择干燥剂须遵循不能与有机溶液发生反应的原则，含金属镁、钙等离子的干燥剂则要注意它们与有机酸，甚至是醇、酚、胺等进行络合的可能性。

2.8.1.2 试剂的无水处理

无水处理是通过物理或者化学方法，将试剂所含水分严格控制在一定的标准，以满足反应要求的深度干燥。

(1) 物理方法

① 共沸蒸馏　很多有机物与水不能混溶，如烃类、卤代烃类、酯类、醚类、高级醇类、酮类等，利用它们可以与水发生共沸的特点，通过蒸馏可除去少量的共沸物——往往表现为乳白色的乳化液。为了得到干的液体组分，往往要待蒸馏出 10% 体积的液体后方可收集，得到干燥有机物。也就是说规范的蒸馏干燥操作是将一无明显分层的有机液体进行常压蒸馏，不仅要将开始馏出的这些乳白色的液体，乃至十分之一总体积的先蒸出的液体都必须作为前馏分。实验室常用试剂如：苯、石油醚、己烷、二氯甲烷、二氯乙烷、四氯化碳、1,4-二氧六环等均可用这种方法进行干燥。本教材中许多有机液体化合物的制备过程中，带有除水性质的蒸馏是经常使用的，也是应该引起初学者注意的一项操作，如乙酸乙酯、乙醚、叔丁基氯等实验中最后的蒸馏过程。需要指出：蒸出十分之一的液体只是在对反应试剂或溶剂的无水处理时才有必要，而对于最终产品的提纯，仅需蒸出混浊液后就可收集主馏分。

事实上，对无水要求较高的实验或生产，往往使用共沸蒸馏带水的方法，以对整个反应体系进行无水处理。如芳香氯的氟代置换，体系中含水量高，反应速度减慢、副产物增加，所以氟化前加入高沸点溶剂（如环丁砜、DMF 等），反应底物后，可用甲苯或二甲苯等进行回流带水，再加入 KF 升温进行氟化。

② 分子筛干燥　一些试剂如甲醇、乙醇低级醇，以及四氢呋喃等，可以与水无限混溶，蒸馏时可与水形成恒沸物。这些试剂中的含水量往往对实验结果产生很大影响，如四氢呋喃是格氏反应中经常使用的溶剂，其含水量若超过 $0.5mL \cdot L^{-1}$，反应将无法引发。在实验室中对这种类型的试剂进行无水处理时使用最为普遍，另一非常有效的手段是分子筛干燥。

分子筛是碱金属硅铝酸盐，其晶格孔道只有几埃（1nm=10Å）。常见规格有 3A、4A、5A 等。3A 分子筛是所有分子筛当中孔道最小的分子筛，其晶格孔道只有 3Å，能吸附临界直径不大于 3Å 的分子；4A 分子筛能吸附临界直径不大于 4Å 的分子，如：水分子、含碳原子数不超过 2 的烃类、H_2S、NH_3、CO_2、O_2、N_2 等小分子。4A 型分子筛也是实验室中最常用的分子筛。水分子极性很强，临界直径为 2.8Å，分子筛中存在大量羟基以及氧原子，对质子性的小分子尤其具有强的亲和性。

分子筛干燥过程包括分子筛的预处理和试剂干燥。将分子筛预先在高温（150~300℃）下马弗炉中灼烧 5~12h，干燥器中置冷后，将分子筛加入到待干试剂中，封好，一般在干燥器中储存 24h 可直接倾出使用。没有马弗炉可用其他加热方法代替，如将分子筛置入圆底烧瓶中，用电热套或硅油浴加热。

分子筛干燥的方法具有很多优点：适用范围广，不仅适用于水溶性试剂，对非水溶性的试剂如酮类、腈、卤代烃、醚、烃类同样适用，不仅适用于液体的无水处理，同样可用于气体的处理；干燥强度相当高，将含水量为 0.5% 的乙醇，通过一装有经过预处理的 4A 分子筛柱，其含水量可降低到 $10\mu L \cdot L^{-1}$ 以下；将试剂中放入分子筛，处理的试剂可以较长期储存（1~2 周），使用方便；分子筛可以回收再使用。

（2）化学方法

金属钠、氢化物（NaH、CaH_2）以及一些金属或非金属的氧化物（CaO、P_2O_5）等也可用于对试剂进行无水处理。这些物质遇水发生不可逆的转化，钠干法是将金属钠切成小块加入到甲醇等试剂中，在干燥氮气氛围形成的无水系统中加热回流至二苯甲酮变色现象出现，蒸馏收集得到干燥甲醇等试剂。钠干法主要用于低级醇（如甲醇、乙醇）、THF 等试剂的无水处理，试剂使用前需要蒸馏。使用钠干法时需要注意的安全性问题是对含水量较大的醇进行处理前可先使用一般干燥进行预处理。氧化钙常用于低级醇、有机胺的无水处理，使用前先将其于 700~900℃ 下灼烧数小时，直接加入到试剂中，使用前无需蒸馏。五氧化二磷往往不直接加入到液体试剂中，而采取与之共存放于干燥器中一段时间至样品恒重。

需要特别指出的是，有机试剂的含水量有时会对实验结果产生致命的影响，所以对其进行无水处理时必须严格按照规范的操作程序进行。常见试剂的无水处理请参考《Purification of Laboratory Chemicals》(Edited by D.D.Perrin；W.L.F.Armarego and D.R.Perrin)。该书中不仅对各种常见试剂的纯化方法作出了系统描述，并对实验室相关操作也有明确的规范。当然无水处理方法可根据试剂的不同性质、实验的不同要求以及实验室具体的条件而灵活地具体确定。

2.8.2　固体的干燥

固体的干燥过程一般是指脱除固体物料所结合的水分，干燥一般须达到恒重。

干燥过程是固体产品最终处理的必经过程，对产品质量有着非常重要的意义。干燥过程也是一种非常复杂的传质、传热过程，其基本原理部分可参阅《现代干燥技术（第二版）》（潘永康等主编，2007年，北京：化学工业出版社）。这里仅介绍几种实验室和工业生产中常用的固体干燥方法。

(1) 直接加热法

最常见的情况是通过给混合物料足够的热能，使水分发生汽化，这就是直接加热法。其适用对象是熔点较高而又具有一定热稳定性的固体。干燥不够，产品含量不能达到标准；过分干燥将导致有机物的质量下降和能量浪费。

实验室加热可采用红外灯加热、电热干燥箱（烘箱）加热。红外灯加热难以量化地控制温度，只能依靠调节灯泡与药品的距离粗略控制。

烘箱一般用于干燥无腐蚀、不挥发的热稳定物质。对用丙酮、乙醇、乙酸乙酯等易燃溶剂进行重结晶的药品，在干燥前以水淋洗，或者在空气中将溶剂全部挥发后方可进入烘箱。烘箱一般附有鼓风系统，这样可及时将湿分带走，加速干燥过程。在具体的操作中，干燥的前奏——过滤过程应很充分，尽量分出自由水分，这样才能保证干燥过程较低的能耗和过程的可重复性。无论使用何种干燥方法，都应遵循彻底过滤的原则。

(2) 空气晾干法

对于热不稳定或低熔点药品，可以采用空气晾干法进行处理，但往往耗时较长，且受到空气流通程度、湿度影响，操作条件不易控制。

(3) 真空干燥技术

对于热、空气不稳定或低熔点而不易挥发的药品，可以辅以真空系统，在减压条件下进行干燥。如果有一定的热稳定性，可同时辅以加热手段来加速药品脱水。实验室中最简单的减压干燥装置如图 2-20 所示，为了缩短干燥时间，样品的预处理即过滤时，须尽量脱除游离水分。对热、氧气不稳定的有机物溶液，普通的蒸馏或挥发是不可取的，用这种简易的干燥装置还可完成溶剂的蒸发，即使处理如二氯甲烷等低沸点溶剂构成的溶液，使用液氮冷阱也可使真空系统免受损害。

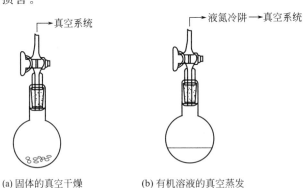

图 2-20 简易减压干燥装置

(a) 固体的真空干燥　　(b) 有机溶液的真空蒸发

也可使用真空烘箱或真空干燥枪来干燥。

生物制品的最终处理广泛使用冷冻干燥技术。冷冻干燥技术最早于 1813 年由英国人 Wollaston 发明，1909 年 Shsckell 试验用该方法对抗毒素、菌种、狂犬病毒及其他生物制品进行冻干保存，取得了较好效果。在第二次世界大战中，对血液制品的大量需求大大刺激了冷冻干燥技术的发展，从此该技术进入了工业应用阶段。冷冻干燥是指把药品溶液在低温

（-60～-40℃）下冻结，然后在真空条件下使水升华，除去冰晶，待升华结束后再进行解吸干燥，除去部分结合水的干燥方法。药品按上述方法冻干后，可在室温下避光长期储存，需要使用时，加蒸馏水或生理盐水制成悬浮液，即可恢复到冻干前的状态。与其他干燥方法相比，药品冷冻干燥法具有非常突出的优点和特点：低温干燥，能使被干燥药品中的热敏物质得以保留；低压干燥，被干燥药品不易氧化变质，同时能因缺氧而灭菌或抑制某些细菌的活力；冻结时被干燥药品可形成"骨架"，干燥后能保持原形，形成多孔结构而且颜色基本不变；复水性好，冻干药品可迅速吸水还原成冻干前的状态；脱水彻底，可长期保存。

药品冷冻干燥机主要由干燥箱、真空系统、制冷系统、冷阱系统、自动控制系统等几大部分组成。此外，大中型冷冻干燥机还常有蒸气灭菌系统（SIP）、在位清洗系统（CIP）。随着生物技术的高速发展，多肽蛋白质类药物不断涌现，可应用于临床的多肽、蛋白酶、激素、疫苗、细胞生长因子及单克隆抗体等成为开发重点，为防止药品变性，目前广泛采用冷冻干燥法制备固态药品。虽然冷冻干燥具有上述优点，但是干燥速率低、干燥时间长、干燥过程能耗高和干燥设备投资大等仍是该技术的缺点。

（4）化学干燥法

对于热不稳定物质，可将之置于敞口的容器中，然后在放有五氧化二磷或氯化钙、浓硫酸、氧化钙等的干燥器中存放至恒重。

附：含水量的测定方法

（1）直接测量

根据固体的性质，选用合适的干燥方法将药品干燥至恒重，干燥前后的质量差即为含湿量。这种方法往往与真实的含湿量有一定差距（干燥方法本身的局限性）。

（2）间接测量法——卡尔-费休法

$$I_2 + SO_2 + 2H_2O \longrightarrow H_2SO_4 + 2HI$$

定量称取含湿药品，用已标定的卡尔-费休试剂（溶解在甲醇中的二氧化硫、碘、吡啶混合物）进行滴定，游离碘的特殊棕色出现时为滴定终点，根据反应式可计算出药品含水量。

其他如色谱法、电导法、红外光谱法等也可以测量某些特定试剂或体系中的水分，在此不作赘述。

2.9 升　华

升华是纯化固体有机化合物的一种方法，它是由固体有机化合物受热而不经熔化直接气化为蒸气，然后由蒸气直接冷凝为固体的过程。固体化合物的蒸气压和固体化合物表面所受压力相等时的温度，称为该物质的升华点。

图 2-21 是物质三相平衡曲线图，ST 表示固相与气相平衡时，固体的蒸气压曲线；TW 是液相与气相平衡时，液体的蒸气压曲线；TV 是固相与液相平衡时的温度和压力关系。三条曲线交叉点 T 是三相点，在这一温度和压力下，固、液、气三相处于平衡状态。

在三相点以下，化合物只有气、固两相，若温度降低，蒸气就不再经过液态而直接变为固体，所以，一般的升华操作在

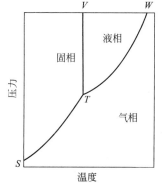

图 2-21　物质的三相平衡曲线

三相点以下的温度进行。若某化合物在三相点以下的蒸气压很高，则气化速度很大，这样就很容易从固体直接变为蒸气。而且，此化合物的蒸气压随温度降低而下降，温度稍微降低，即可由蒸气直接变为固体，这种化合物在常压下很容易用升华方法来纯化。

常压下蒸气压过低不易升华的物质，若放在减压条件下升华，便可得到满意的效果。也可采用在减压下通入少量空气或惰性气体的方法来加快升华的速度。

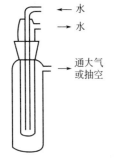

图 2-22 常（减）压升华装置

2.9.1 萘的常压升华

称取 0.05g 萘，放在干燥的升华管中，插入内管（见图 2-22），通入冷凝水，用电热套加热，控制温度在 75～80℃。萘慢慢升华，几分钟后，在内管外壁底部长满片状晶体。停止加热，关闭冷凝水，取出内管，将其底部的萘用小刮刀刮下，测其熔点，并称重。

2.9.2 萘的减压升华

称取 0.05g 萘，放入干燥的升华管中，插上内管，开通冷凝水，外管的支管接水泵减压。把升华管放在 70℃ 左右的水浴中加热。几分钟后，停止加热和减压，关闭冷凝水，取出内管，将其底部的萘用刮刀刮下，测其熔点，并称重。纯萘：m.p. 为 80.55℃。

思 考 题

(1) 升华操作时，为什么要缓缓加热？如升温过高有什么坏处？
(2) 熔点与三相点有何不同？

2.10 色 谱

色谱是分离、提纯和鉴定有机化合物的重要方法之一。色谱法的分离效果远比分馏、重结晶等一般方法好，特别适用于少量（微量）物质的处理，在化学、化工、生物、医药中得到了广泛的应用。

色谱的基本原理是利用混合物各组分在某一物质中的吸附或溶解性能（分配）的不同，或其他亲和作用的差异，使混合物流经该种物质，进行反复的吸附或分配等作用，从而将各组分分开。因此其有两个相：固定相和流动相，根据组分与固定相的作用原理不同，可将色谱法分为吸附色谱、分配色谱、离子交换色谱等。根据操作条件的不同可分为：柱色谱、纸色谱、薄层色谱、气相色谱和高效液相色谱等。

2.10.1 柱色谱

柱色谱（又称为柱上层析）常用的有吸附柱色谱和分配柱色谱。前者常用氧化铝和硅胶作固定相，后者则是以硅藻土、纤维素等为支持剂，以吸收的液体为固定相，支持剂本身不起分离作用。

吸附柱色谱（如图 2-23 所示）通常是在玻璃管中装入一定的固体吸附剂，被分离物则被吸附在柱的顶端，当洗脱剂流下时，由于不同化合物的吸附能力不同，便以不同的速度下移，于是形成了不同的层次，而将混合物分离开来。

(1) 吸附剂

常用的吸附剂有氧化铝、硅胶、活性炭等。其颗粒大小要均匀，一般说来，颗粒越小，

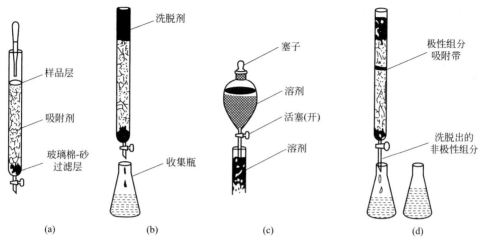

图 2-23 色谱柱装置与分离过程

表面积越大,吸附能力越高,但是洗脱剂的流速越小。因此颗粒大小应该根据具体分离需要来确定。大多数吸附剂都能强烈地吸附水,而且水不易被其他化合物置换,因此使吸附剂的活性降低。通常用加热的方法使吸附剂活化。

(2) 溶质的结构与吸附能力

化合物的吸附能力与它们的极性成正比,即化合物中含有极性较大的基团,其吸附能力也较强。氧化铝对有关物质的吸附能力按以下次序递减:

酸和碱＞醇、胺、硫醇＞酯、醛、酮＞芳香族化合物＞卤代物、醚＞烯＞饱和烃

(3) 溶剂（又称为洗脱剂）

洗脱剂的选择是关系分离效果很重要的方面。通常根据被分离物中各组分的极性、溶解度和吸附剂的活性来考虑。一般洗脱剂的极性大,洗脱能力大。常用的洗脱剂的极性由小到大的次序是:

己烷、石油醚＜四氯化碳＜三氯乙烯＜二硫化碳＜甲苯＜苯＜二氯甲烷＜氯仿＜乙醚＜乙酸乙酯＜丙酮＜乙醇＜甲醇＜水＜吡啶＜乙酸

具体进行柱色谱时分以下步骤。

① 装柱 可以采用湿法和干法。目前一般采用下有砂芯的玻璃管作色谱柱,湿法装柱方法如下:向柱中加入 3/4 的洗脱剂,打开活塞,控制流出速度为1滴/秒。通过一个干燥的玻璃漏斗慢慢加入色谱用氧化铝或硅胶,同时用木棒轻轻敲打柱身,当装柱至 3/4 时,再加一层石英砂或一小片滤纸。操作和下面装样品、洗脱时均不能使液面低于沙子或滤纸的上层。所装的色谱柱要求填装均匀,不能有裂缝或气泡。

② 装样品 首先将样品溶解于一定的溶剂中,溶剂的极性尽量小,体积尽量小。当洗脱剂刚好流至石英砂或滤纸的上面时,将样品溶液沿柱壁加入,当此液面将至石英砂（或滤纸）面时,立即加入少量的洗脱剂洗下管壁上的被分离的物质,如此 2~3 次。

③ 洗脱 在装好样品的色谱柱上装置滴液漏斗,其中倒入洗脱剂,控制流出速度为1滴/秒,进行洗脱。对于复杂体系一般可进行梯度洗脱,即慢慢改变洗脱剂的极性（由小到大）。收集洗脱液时简单的方法是根据颜色的不同来改变接收瓶,一般是每接收一定的体积更换接收瓶,然后根据薄层色谱来确定样品组成。

④ 浓缩收集样品 根据薄层色谱结果将相同组成的样品并在一起进行蒸馏（或减压蒸馏）蒸除洗脱剂，当快蒸干时，转移至蒸发皿中，在水浴上蒸干或用红外灯烘干。

2.10.2 薄层色谱

薄层色谱是一种微量、快速而简单的色谱法，它兼备了柱色谱和纸色谱的优点。特别适用于挥发性较小，或在较高温度下易发生变化而不能用气相色谱分析的物质。薄层色谱也有吸附色谱和分配色谱两种，同样分别用氧化铝或硅藻土、纤维素。

吸附剂常用的也是氧化铝和硅胶，与柱色谱不同的是，这里常用的是硅胶 H——不含黏合剂，硅胶 G——含有煅石膏作黏合剂，硅胶 HF_{254}——含荧光物质，可在 254nm 的紫外光下观察荧光，硅胶 GF_{254}——含有煅石膏和荧光物质。同样氧化铝也有相应的品种。

薄层色谱的分离原理与柱色谱类似，其操作步骤如下。

① 薄层板的制备 将吸附剂和水调成糊状，然后在洗干净的平整的玻璃板上铺成均匀的薄层，再把薄层板放在已校正水平的平台上晾干，最后将干透的薄层板置于烘箱中加热活化，活化时需要慢慢升温。硅胶板在110℃左右活化 30～60min，可得Ⅳ～Ⅴ级活性的薄层板；氧化铝薄层板在 200～220℃时烘 4h，可得Ⅱ级活性的薄层板；在 150～160℃时烘 4h，可得Ⅲ～Ⅴ级活性的薄层板。所制得的薄板应该均匀、没有裂缝。将符合要求并活化的薄板置于干燥器中保存备用。

② 点样 在薄层板的一端约 1cm 处，用铅笔轻轻划一条线。将样品用低沸点溶剂配成 1% 溶液，用管口平整且内径小于 1mm 的毛细管吸取样品溶液，轻轻接触起点线，如果溶液太稀，可待溶剂挥发后重复点样。点样斑直径一般不超过 0.3cm，若多处点样，点样距离为 1cm 左右。

③ 展开 薄层色谱展开剂的选择与柱色谱一样，主要根据样品的极性、溶解度和吸附剂的活性来考虑。展开剂的极性越大对化合物的洗脱力越大。展开需要在封闭容器中进行，将展开剂放在展开槽中，使溶剂蒸气饱和 5～10min，将点好样的薄层板垂直或倾斜放入展开槽中展开（如图 2-24 所示）。当展开剂前沿上升到离薄层板顶端 1cm 处时取出薄层板，用铅笔或小针划出溶剂前沿，放平晾干。

④ 显色 如果用的是有荧光的氧化铝或硅胶，可在紫外光下观察，并标出斑点；或将薄层板置于碘缸中显色。根据斑点位置计算比移值 R_f（如图 2-25 所示）。

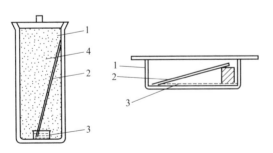

图 2-24 薄层色谱展开示意图
1—色谱缸；2—薄层板；3—小皿盛展开剂；4—展开剂蒸汽

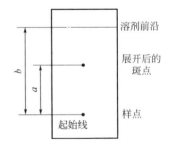

图 2-25 薄层色谱比移值的确定

$$R_f = 溶质所经过的距离/溶剂所经过的距离 = \frac{a}{b}$$

R_f 随化合物结构、固定相与流动相的性质、温度等因素的不同而变化。当温度等实验条件固定时,比移值是一个常数,可作为定性分析的依据。但是由于影响 R_f 的因素很多,实验数据与文献记载往往不完全相同,因此在鉴定时常常采用标准样品对照分析。

2.10.3 气相色谱

气相色谱(gas chromatography,简称 GC)是 20 世纪 50 年代发展起来的一种分离分析技术,是以气体为流动相的色谱法,根据固定相的不同可分为气-固色谱和气-液色谱两类。气-固色谱的固定相通常用的是硅胶、氧化铝和分子筛等,主要是利用不同组分在固定相表面吸附能力的不同而达到分离的目的,因此属于吸附色谱。而气-液色谱的固定相是在小颗粒固体(载体)表面吸附的高沸点液体(固定液),它利用被分析样品各组分在固定液中的溶解度差异而将混合物分离,属于分配色谱的一种形式,具有快速、高效、高灵敏度的特点,适用于沸点在 300℃ 以下、热稳定的挥发性物质的分离和测定。目前,该有关仪器已经成为大多数有机合成实验室和工厂进行中控、产品分析的常用仪器,有关知识在仪器分析和有机分析等课程和教材中有详细介绍,这里仅对一些基础知识和简单的使用方法作介绍。

(1) 气相色谱流程

常用的气相色谱仪由色谱柱、检测器、气流控制系统、温度控制系统、进样系统和信号记录系统等设备所组成,其流程如图 2-26 所示。

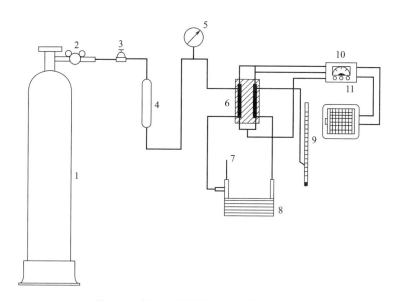

图 2-26 气相色谱流程及仪器设备示意图
1—载气钢瓶;2—减压阀;3—流量调节阀;4—净化器;5—压力表;6—检测器;
7—进样器和气化室;8—色谱柱;9—流量计;10—测量电桥;11—信号记录仪

(2) 基本原理

如上所述,样品在气化室气化,随着载气进入色谱柱,各组分因分配比的不同或吸附能力的不同,而逐渐分开,进入检测器,便可得到色谱图,依据出峰时间(保留值,如图 2-27 所示)和峰面积可进行定性和定量分析。

(3) 操作步骤

按照说明书正确安装仪器,并按照有关教材安装和制备色谱柱。首先按照说明书对仪器

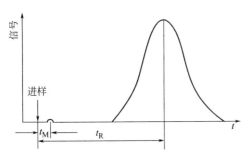

图 2-27 保留值示意图
t_R—从进样到色谱出峰最高点的时间

进行调试（如载气、氢气、空气等的流量，灵敏度，量程），选择柱温、气化和检测温度等。并使仪器稳定一段时间。用微量注射器吸取一定体积的样品（固体样品可用低沸点的溶剂溶解，配成溶液），从进样器进样，同时开动信号记录仪记录。实验结束时首先关闭气化室、柱和检测器的加热电源，并关闭氢气和空气，待有关温度冷却到接近室温时关闭载气。

2.10.4 高效液相色谱

高效液相色谱（high performance liquid chromatography，HPLC）也称高压液相色谱（high pressure liquid chromatography，HPLC），是 20 世纪 60 年代后期迅速发展起来的快速、高效、高灵敏度的分离方法，而且对一些高沸点、热稳定性差、离子型等气相色谱难以进行分离分析的化合物特别适用，因此高效液相色谱得到了广泛的应用。在高效液相色谱中按照固定相和流动相相对极性的不同分正相和反相系统两种，分离过程中根据机理的不同又有以下五种。

① 液固吸附色谱　组分按其在两相吸附作用的大小进行分离。

② 液液分配色谱　固定相的液体承载在载体上，流动相和固定相的液液是互不相溶的，按组分在两相中溶解度的不同而进行分离。

③ 离子交换色谱　离子交换剂作固定相，不同组分按其离子交换能力的不同进行分离，是分离离子型化合物较好的方法。

④ 凝胶渗透色谱　又称排阻色谱，按分子量大小不同进行分离是凝胶渗透色谱的特点，其在分离分析中起着十分重要的作用。

⑤ 亲和色谱　亲和色谱又称生物亲和色谱，利用组分对固定相亲和力的差别进行分离，是蛋白质、酶等分离提纯的有效方法。

高效液相色谱的流程如图 2-28 所示，有关操作注意事项见有关仪器的说明书和有关专著。

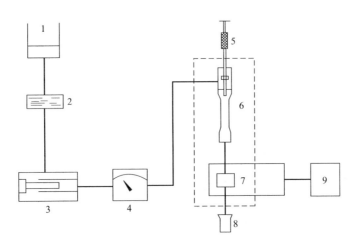

图 2-28 高效液相色谱流程示意图

1—贮液罐；2—过滤器；3—高压泵；4—压力表；5—进样器；6—色谱柱；7—检测器；8—样品收集；9—记录仪

2.11 波 谱 技 术

随着科学技术的发展，红外光谱（IR）、紫外光谱（UV）、核磁共振谱（NMR）和质谱（MS）等波谱技术已经成为测定有机化合物结构的主要手段。本书对有关内容作简单的介绍，主要是介绍样品的准备和谱图的解析，以达到为有机化学实验服务的目的。

2.11.1 红外光谱（IR）

光是一种电磁波，红外光是介于可见光和微波之间的电磁波。对有机化合物而言，重要的是 $2.5 \sim 15 \mu m$（即 $4000 \sim 660 cm^{-1}$）的红外光。

分子的运动有平动、转动和振动，而分子中存在着两种基本的振动：伸缩振动（键长发生改变的振动）和弯曲振动（键角发生改变的振动）。各个化学键的振动频率不仅与化学键本身有关，而且受到整个分子的影响。当一定频率的红外光照射分子时，如果分子的一个振动频率与红外光的频率相同，红外光则被吸收。这样连续改变红外光的频率就可以观察到红外吸收光谱（频率或波长为横坐标，吸收率或透射百分率为纵坐标），图 2-29 为甲苯和正己烷的红外光谱图。一般红外光谱图分为两个主要区域：官能团区为 $2.5 \sim 7.5 \mu m$（$4000 \sim 1300 cm^{-1}$）；指纹区为 $7.7 \sim 15 \mu m$（$1300 \sim 660 cm^{-1}$）。前一个区域一般是由两个原子振动所产生的，与整个分子的关系不大，这个区域对判别化合物的官能团起着重要作用。指纹区的基本振动是多原子体系的伸缩和弯曲振动，与整个分子有关，这个区域中每个化合物有着彼此不同的谱图，就像人们的指纹一样，没有两个指纹是相同的，所以这个区域称为指纹区。指纹区对鉴定两个化合物是否相同有着很大的作用。

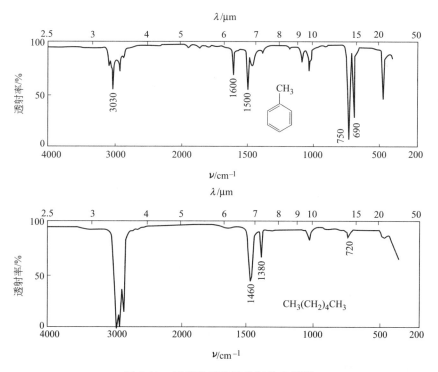

图 2-29 甲苯和正己烷的红外光谱图

由此可见，分子的振动频率大小决定了红外光谱横坐标的位置，而分子的振动有多种形式，影响因素也有很多。对双原子的振动频率（如图 2-30 所示）可用下式表示：

$$\nu = \frac{1}{2\pi c}\sqrt{\frac{k}{\mu}}$$

$$\mu = \frac{m_1 m_2}{m_1 + m_2}$$

图 2-30 双原子的振动示意图

式中，ν 为振动带的波数；c 为光速；k 为键的力常数；μ 为所涉及原子的折合质量；m_1 为第一个原子的质量；m_2 为第二个原子的质量。

由上述公式可知：原子质量的增加使吸收频率减小，Y—H 类型的官能团（Y=O、N、C、卤素等）的吸收频率最大。键的力常数增加吸收频率增大，如碳碳叁键的伸缩振动频率大于碳碳双键的伸缩振动频率。各官能团的特征吸收谱带位置见附录 6。

进行红外光谱分析的样品一般要求是纯样品，不能含有水，气体和液体样品用相应的池子装样。固体样品一般与溴化钠压片，也可与石蜡油或氟油（Fluorolub）等混合成糊状，再将这种糊状物夹在两块盐片之间形成一种半透明的薄膜进行测定。

2.11.2　紫外光谱（UV）

波长在 200～400nm 的电磁波是紫外光，波长在 400～800nm 的电磁波是可见光，如果紫外光或可见光照射到含有共轭的不饱和的化合物溶液中，这时可以看到一部分光线被吸收了。如果以波长为横坐标，以吸收强度 A 为纵坐标，则得到紫外可见光谱（如图 2-31 所示）。

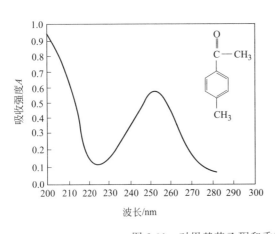
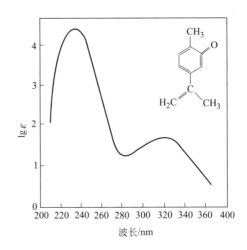

图 2-31　对甲基苯乙酮和香芹酮的紫外吸收光谱图

吸收强度 A 与入射光强度 I_0、透射光强度 I 之间有如下关系：$A = \lg(I_0/I)$。而摩尔吸光度与吸收强度 A 之间的关系如下：

$$\varepsilon = \frac{A}{cl}$$

式中，c 为溶液的物质的量浓度；l 为吸收池的厚度，cm。

根据这一关系可以测定化合物的纯度。

在化学文献中一般用以下形式报告紫外光谱数据：如甲基苯乙酮 $\lambda_{\max}^{CH_3OH}$ 252nm，$\varepsilon = 12300$，或 $\lambda_{\max}^{CH_3OH}$ 252nm$(\lg\varepsilon = 4.09)$。

分子对紫外光和可见光的吸收取决于分子的电子结构（电子能级），对于一般饱和化合物而言，只有σ电子和n电子，其电子能级的跃迁主要是σ→σ*和n→σ*，这些吸收都不发生在近紫外区。对于含有孤立双键的化合物，有σ、n和π电子，其能级最低的跃迁是π→π*和n→π*，这些吸收绝大多数也不在近紫外光区，并且吸收非常弱。因此紫外光谱绝大部分仅限于共轭体系。例如乙烯的最大吸收波长在171nm处，1,3-丁二烯的最大吸收波长则在217nm(ε＝21000)处，2-丁烯醛的最大吸收波长在220nm(ε＝15000)，分子中含有的共轭双键越多，最大吸收波长就越长。β-胡萝卜素是一个有11个共轭双键的化合物，它的最大吸收波长在497nm处。一些不饱和化合物和共轭不饱和化合物的最大吸收波长见表2-6。

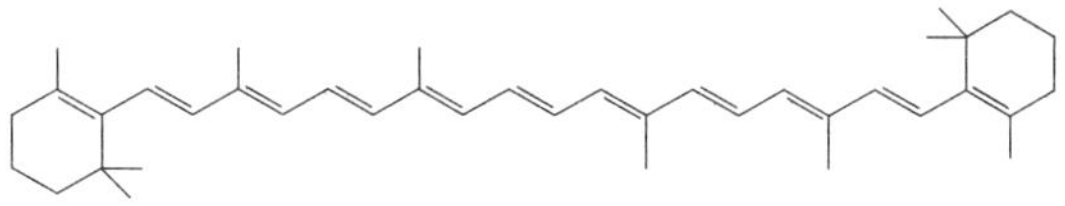

β-胡萝卜素

表2-6 一些不饱和化合物和共轭不饱和化合物的最大吸收波长

化合物	结构	λ_{max}/nm	ε_{max}	λ_{max}/nm	ε_{max}
乙烯	$CH_2\!=\!CH_2$	171	15530		
1-辛烯	$CH_3(CH_2)_5CH\!=\!CH_2$	177	12600		
1-辛炔	$CH_3(CH_2)_5C\!\equiv\!CH$	185	2000		
1,3-丁二烯	$CH_2\!=\!CH\!-\!CH\!=\!CH_2$	217	21000		
1-丁烯-3-炔	$CH_2\!=\!CH\!-\!C\!\equiv\!CH$	228	7800		
1,4-戊二烯	$CH_2\!=\!CHCH_2CH\!=\!CH_2$	178			
环己烯		182	7600		
1,3-环戊二烯		239	3400		
1,3-环己二烯		256	8000		
乙醛	CH_3CHO			290	16
丙酮	$CH_3\!-\!\overset{O}{\underset{\|}{C}}\!-\!CH_3$	188	900	279	15
丙烯醛	$CH_2\!=\!CH\!-\!CHO$	210	11400	315	26
2-丁烯醛	$CH_3\!-\!CH\!=\!CH\!-\!CHO$	220	15000	322	25
甲基乙烯基酮	$CH_3\!-\!\overset{O}{\underset{\|}{C}}\!-\!CH\!=\!CH_2$	212.5	7100	320	27

对芳香族化合物，苯和烷基苯在紫外区域有两个吸收带：E带，波长在200nm左右，吸收强；B带，波长在260nm附近，吸收弱，有精细结构，许多芳香族化合物都有这样的吸收带。乙醇中苯、萘、蒽的紫外吸收光谱如图2-32所示。

2.11.3 核磁共振谱（NMR）

对自旋量子数 I 为1/2的核R如 $_1^1H$、$_6^{13}C$、$_9^{19}F$、$_{15}^{31}P$、$_7^{15}N$ 等，在外加磁场 H_0 中有两

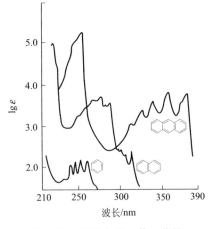

图 2-32　乙醇中苯、萘、蒽的紫外吸收光谱

个取向，具有不同的能量。若外加适当频率 ν 的电磁辐射，就可能导致核跃迁的发生，上述电磁辐射的频率为 ν：

$$\nu = \frac{\gamma H_0}{2\pi}$$

式中，γ 为旋磁比或磁旋比，为基本常数，代表有关核的特性。

根据上式，若保持磁场强度不变，改变射频（称为扫频，相反称为扫场），直至磁场强度和射频的组合达到有关核的特征要求，就能观察到核的跃迁。这种情况通常称为共振。

(1) 核磁共振谱仪

一般的核磁共振谱仪的结构如图 2-33 所示。主要包括磁铁、扫描发生器、射频发生器、接收器线圈和记录仪等。

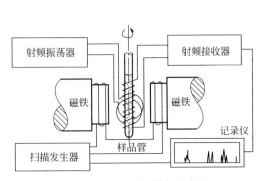

图 2-33　核磁共振谱仪示意图

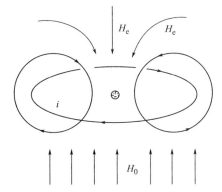

图 2-34　氢原子周围电子对质子的屏蔽作用

(2) 化学位移

虽然从上式得到相同的核在同一磁场强度或同一频率下发生核磁共振，但实际上并不如此。因为共振频率不完全取决于外加磁场 H_0，而是取决于作用在共振核上的实际磁场值的大小。实际上，在质子的周围还存在着运动的电子等影响，这些电子产生了屏蔽作用（如图 2-34 所示）。此外，苯环、双键、叁键的 π 电子对周围的质子也可产生屏蔽作用（或去屏蔽作用），这些称为磁的各向异性（如图 2-35 所示）。因此由于分子中电子云密度的不同和磁的各向异性的不同使得不同环境的质子在不同的频率发生核磁共振，即产生了化学位移。

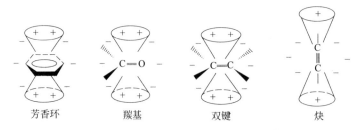

图 2-35　与磁的各向异性基团有关的屏蔽区

化学位移的定义是标准化合物的共振频率与某一个质子的共振频率之差。由于该差值相对于共振频率而言非常小，因此常常用相对值来表示化学位移，通常用符号 δ 来表示，一般的核磁共振谱图的横坐标是化学位移（由大到小），即磁场是从小到大；纵坐标为吸收强度（如图2-36所示）。

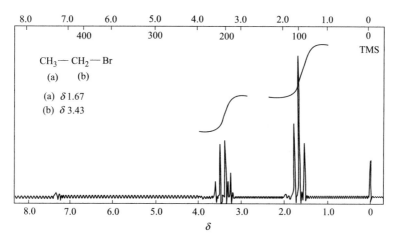

图2-36　溴乙烷的核磁共振谱图

常用的标准化合物是四甲基硅烷（tetramethylsilane，简称 TMS）。一些常见基团质子的化学位移值见附录7。

（3）峰面积

在核磁共振谱图中，每组峰的面积与产生这组信号的质子数目成正比，如果把各组信号的面积进行比较，就能确定各种类型质子的相对数目。目前所使用的核磁共振谱仪一般将每个吸收峰的面积进行了积分，并在谱图上记录了积分曲线。

（4）自旋-自旋偶合——峰的裂分

从溴乙烷的核磁共振谱图上可以看到不是两个单峰，而是两组峰。这种裂分现象是由于质子与邻近质子相互作用的结果，称为自旋-自旋偶合。一般说来，对某一个质子来说只有在和它直接相邻的碳上有非等性质子时才产生自旋-自旋偶合（当然同一碳上的两个非等性质子之间也可以表现出偶合）。所谓非等性质子是指电子环境不同的质子，也就是在核磁共振谱图上化学位移不同的质子。一般而言，裂分数与相邻质子数有关，即裂分数为 $n+1$，n 是相邻质子的数目。

综上所述，从有机化合物的核磁共振谱可以得到大量的结构信息：质子的类型（化学环境）、质子的数目和相邻质子的数目等。

2.11.4　质谱（MS）

质谱是在20世纪40年代发现的，直到70年代才在有机化学、生物化学等领域得到广泛使用，目前已经成为不可缺少的工具。

质谱仪示意如图2-37所示，包括三个主要的组成部分：离子源、分析系统和离子收集、检定系统。图2-38是一张质谱图，横坐标是质荷比 m/z，纵坐标是离子的相对丰度。

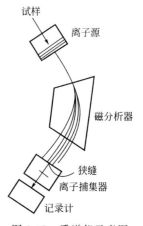

图2-37　质谱仪示意图

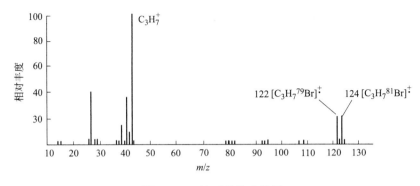

图 2-38 1-溴丙烷的质谱图

质谱法在有机化学实验中有多种用途，主要是测定相对分子质量、分子式和有机化合物的结构等。相对分子质量的确定主要是根据分子离子峰来确定；分子式的确定主要可根据相对分子质量和贝农表来确定；结构可根据碎片峰和裂解方式来确定。有关详细内容见有关专著。

2.12　Schlenk Technique and Glove Box Operation

Schlenk technique is one of the most important organic synthesis method. This method supplies a dry and air-free circumstance to handle the oxygen and moisture sensitive organometallic compounds in the synthetic and post-treatment processes. This operation method is simple, but not easy to operate. Recently, glove box has been developed to replace the Schlenk technique. Glove box is easy to supply a dry and air-free circumstance to store and operate the air and moisture sensitive compounds. In this part, we present a brief introduction of Schlenk technique and glove box operation

2.12.1　Schlenk Technique

2.12.1.1　Schlenk line

The Schlenk line (also vacuum gas manifold) is a commonly used chemistry apparatus developed by Wilhelm Schlenk. It consists of a dual manifold with several ports. One manifold is connected to a source of purified inert gas, while the other is connected to a vacuum pump. The inert-gas line is vented through an oil bubbler, while solvent vapors and gaseous reaction products are prevented from contaminating the vacuum pump by a liquid-nitrogen or dry-ice/acetone cold trap. Special stopcocks or Teflon taps allow vacuum or inert gas to be selected without the need for placing the sample on a separate line.

Schlenk lines are useful for safely and successfully manipulating air-sensitive compounds. The vacuum is also often used to remove the last traces of solvent from a sample. Vacuum and gas manifolds often have many ports and lines, and with care, it is possible for several reactions or operations to be run simultaneously.

When the reagents are highly susceptible to oxidation, traces of oxygen may pose a problem. Then, for the removal of oxygen below the ppm level, the inert gas needs to be purified by passing it through a deoxygenation catalyst. This is usually a column of copper (Ⅰ) or manganese (Ⅱ) oxide, which reacts with oxygen traces present in the inert gas.

The main techniques associated with the use of a Schlenk line (Fig. 2-39) include:

① Counter flow additions, where air-stable reagents are added to the reaction vessel against a flow of inert gas;

② The use of syringes and rubber septa to transfer liquids and solutions;

③ Cannula transfer, where liquids or solutions of air-sensitive reagents are transferred between different vessels stoppered with septa using a long thin tube known as a cannula. Liquid flow is supported by vacuum or inert-gas pressure.

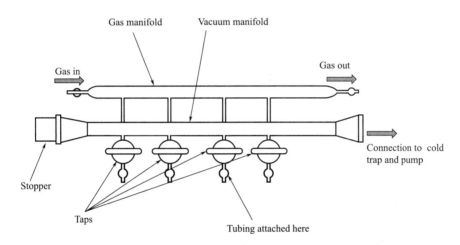

Fig. 2-39 Schlenk line

2.12.1.2 Operation procedure

(1) Starting Up a Schlenk Line

(a) Make sure there is enough liquid nitrogen in the Dewar to fill the trap and to do whatever else you will be doing during the day's lab session. If you empty the Dewar, refill it while you wait for your vacuum pump to warm up.

(b) Make sure all of the stopcocks are properly greased. Note: Use only Apiezon M grease on your Schlenk line.

(c) If the vacuum trap is dry, make sure it is greased and all of the stopcocks are closed and start the vacuum pump. Fill the trap with liquid nitrogen. Begin working.

(d) If the vacuum trap has solvent in it, empty the trap into the waste solvent bottle. Assemble the trap making sure the stopcocks are closed. Fill the trap with liquid nitrogen, wait a moment, then start the pump. Wrap the trap with a towel. *Important: When using this method, liquid oxygen can collect in the trap. It is highly explosive especially when in contact with organic compounds. Never let the trap sit in liquid nitrogen for more than a minute or so without having the vacuum pump turned on.*

(e) Make sure there is sufficient nitrogen in the cylinder to do your work. Turn on the nitrogen flow valve. Crack open the bubbler stopcock and let the line purge gently for 5-10min. Close the stopcock and open the valve until it is fully open. (The nitrogen regulator should never be set higher than 5 psi.)

(2) Shutting Down a Schlenk Line

(a) Make sure all vessels attached to the line are under nitrogen.

(b) Shut off the pump and immediately vent the line through one of the stopcocks.

(c) Drop the trap immediately or liquid oxygen will begin to collect in the trap.

(d) Close the nitrogen feed valve. If you are leaving something open to nitrogen then partially close the feed valve. Open the bubbler stopcock and adjust the flow rate to a gentle bubble, then close the bubbler stopcock.

2.12.1.3 Notes for Schlenk line operation

Liquid oxygen is a light blue colored liquid. It is highly explosive, particularly in the presence of organic compounds such as the grease that we use. If you should ever make liquid oxygen, immediately remove any liquid nitrogen coolant, vent the vessel, and leave the room quickly, closing the door behind you. Put warning notes on the doors and tell someone on the floor what is going on. Remain in line-of-sight of the lab and make sure no one goes in. Wait 15-30min and then check the trap carefully.

2.12.2 Glove Box Operation

2.12.2.1 Glove box

Air-free technique gives a broad overview of methods including:

Glovebox—used to manipulate air-sensitive (oxygen-or moisture-sensitive) chemicals.

Schlenk flask—reaction vessel for handling air-sensitive compounds.

Perkin triangle—used for the distillation of air-sensitive compounds.

The glove box provides an interior chamber for handling substances that are sensitive to oxygen and/or moisture. It is straightforward to use but it is very important to follow the procedures below to maintain the conditions inside the box. Please remove all jewelry from your hands and wrists before using the glove box to avoid ripping the gloves. Make an entry in the log book before you start to use the glove box.

2.12.2.2 Operation procedure

(1) Touch Screen

(a) The system constantly monitors the pressure, moisture and oxygen levels and will alarm if levels fall outside set parameters;

(b) The touch screen displays the current levels, provides access to the box controls and highlights any alarms;

(c) The interior box light is controlled from here under "functions".

(2) Operating the Mini Port

(a) Port should be left under vacuum after use so it will be necessary to fill port with gas from the box by turning the black hand valve on port to "refill" position. When filled return valve to "close" position;

(b) To open door pull quick lock lever and remove cover;

(c) Place items into port, replace cover and push close the quick lock lever (with care not to trap fingers);

(d) Evacuate port by turning hand valve to "evacuate", wait for the pressure to drop below-0.9 mbar;

(e) Refill port by turning hand valve to "refill", wait until pressure returns to normal;

(f) Repeat steps d and e two more times (leaving pressure normal). Then return valve to "close" position;

(g) DO NOT open interior door until port has been through this process THREE times, this prevents interior of box becoming contaminated;

(h) The interior door opens and closes in the same way as the exterior;

(i) Remove all items from port and close door.

(3) Removing items from box

(a) Place all items to be removed into port;

(b) Close interior door and secure shut;

(c) Then simply open exterior door and remove items from inside;

(d) Close door afterwards;

(e) DO NOT open both doors at the same time.

(4) After use

(a) Ensure all exterior and interior port doors are closed, evacuate and refill port and evacuate again to leave under vacuum;

(b) Check valves on circular port are closed;

(c) Check valves on mini port are turned to "close" position;

(d) If you are leaving items inside the box please label clearly with the details (Fig. 2-40).

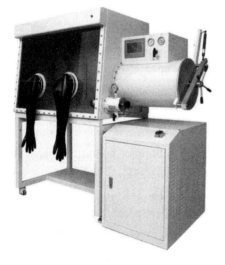 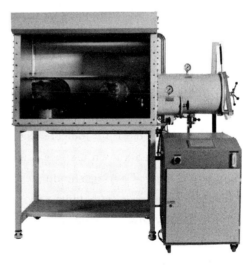

Fig. 2-40　Glove box

2.13　Reaction Monitoring

Most people who are new recruits to the research labs learned their basic skills for carrying out reactions in an undergraduate laboratory. Inevitably, most of the organic chemistry undertaken in these lab classes involves following 'recipes', which have been well tried and tested. Therefore the conditions and the time taken for the reactions to reach completion is well established, and work-up can be carried out after a present time. Unfortunately, the idea that you can guess the time it takes for a reaction to reach completion is a very bad habit to carry over into a research environment.

Every reaction you carry out should be monitored, and one of the first things you should do before starting any reaction is to decide on a suitable method for monitoring its progress. Even if you are following a literature procedure, reaction monitoring is still essential and it will usually save you time as well as giving you confidence about what is happening. Carrying out a reaction without monitoring its progress is like trying to thread a needle with your eyes closed!

The simplest and most universal method of reaction monitoring is thin layer chromatography (TLC) and this will be discussed first of all, but it is not always the best or only method, and sometimes you may have to use a little ingenuity to find an appropriate reaction monitoring technique.

2.13.1　Thin Layer Chromatography（TLC）

TLC is a simple, but extremely powerful analytical tool. However it may take a little time before your expertise reaches a consistently high level, since a certain time of intuition is always involved in choosing the appropriate solvent system, spotting the correct amount of sample, etc. Once you have gained experience and confidence in the use of TLC, you will find it extremely useful for a variety of purposes.

2.13.1.1　The main uses of TLC

（1）TLC is normally the simplest and quickest way to monitor a reaction and the reaction mixture should be chromatographed against starting materials (and a co-spot). This allows you to follow how the reaction is progressing, and to assess when is the best time to work it up. In all cases a record of the TLC should be made in your lab book.

（2）TLC can be used to indicate the identity of a compound, by comparing the unknown sample, with a known material. In general each substance is spotted separately and also together (co-spot). Caution should be applied as co-running on TLC is not definitive proof of identity. Of course, substances that do not co-run are definitely not the same.

（3）TLC usually gives a good indication of the purity of a substance. Diastereoisomer can usually (but not always), be distinguished.

（4）For flash chromatography, TLC is first used to determine the solvent system and quantity of silica required, and secondly to monitor the column fractions.

2.13.1.2 TLC plates

There are two main types of coating for TLC plates, silica and alumina. Silica plates are most commonly used, they are slightly acidic and are suitable for running a broad range of compounds. Most of the information in this section refers directly to silica plates, but the same principles apply when using alumina plates. Alumina plates are slightly basic and are quite commonly used when a basic compound will not run very well on silica.

The most common TLC plates have either a glass or plastic backing coated with a thin layer of silica, which contains a binding agent to keep it bonded to the backing. Although TLC plates can be home-made, most people prefer to use commercial ones as they give very consistent results. For analytical purposes plates with a 0.25mm layer of silica are normally used. They are available in a variety of sizes, although 5cm×20cm is probably the most convenient size. The chromatographs are run along the 5cm length, cut to an appropriate width. This normally gives adequate resolution, with a very short running time. Plastic backed plates are sometimes cheaper and can easily be cut into strips with scissors, but glass plates seem to give better resolution and can be heated more vigorously for visualization purposes.

To cut a glass TLC plate, place it face down on a clean piece of paper, hold a ruler firmly along the proposed cut and draw a sharp diamond glass cutter along the line once only. Then holding the plate with the forefinger and thumb of each hand, on either side of the score line snap plate along the line. With practice, and a good glass cutter, you should be able to cut plates down to about 1.5cm without ever spoiling them.

2.13.1.3 The procedure for running a TLC

(1) Cut a TLC plate which is 5cm long, and wide enough so that about 0.5cm can be left between each spot (obviously the width of the plate is dependent upon the number of spots to be run on it).

(2) Make a TLC spotter by drawing out a Pasteur pipette or melting point tube to about 0.5mm using a micro burner. The spotter can be used many times provided you wash it with clean solvent in between runs.

(3) If the substance to be analyzed is not already in solution, take it up in a volatile solvent as a ca. (1—2)% solution. A non-polar solvent is preferable but methylene chloride is often used as a universal solvent for TLC samples. Reaction mixture solutions can be diluted down if necessary, and with experience making samples to a reasonable strength becomes intuitive.

(4) Using the spotter, spot a small amount of each solution about 0.5cm from the bottom of the plate, leaving a similar distance between each spot. The spots should be kept as small as possible, and you should take care to make sure that each of the spots is the same distance from the bottom of the plate [Fig. 2-41(a)]. The absolute distance which a compound runs up a TLC plate is extremely variable, depending on the exact conditions under which the plate was run. It is therefore much more informative to run comparative TLCs. When analyzing a reaction mixture, it should always be run in comparison with the starting material and a mixed spot should also be run. This is very important because it often

enables you to distinguish between compounds which run in almost identical positions.

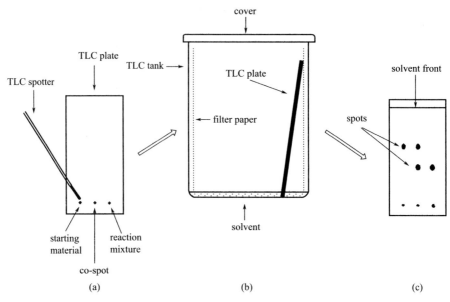

Fig. 2-41 The procedure for running a TLC

(5) Place the plate upright in a tank lined with a filter paper and containing the chosen solvent system [to a depth of ca. 0.3cm as shown in Fig. 2-41(b)]. Allow the solvent to creep up the TLC plate until it is ca. 0.3cm from the top, then remove it and mark the level of the solvent front [Fig. 2-41(c)]. A 150mL lipless beaker makes a convenient TLC tank and a petri dish can be used as a cover.

2.13.1.4 Detecting the spots

The three general ways to visualize spots on TLC plates are listed below. Any one or combination of these techniques can be used, but they should be carried out in the order shown, as the first two techniques are nondestructive.

(1) The plate can be viewed under a ultraviolet lamp to show any UV-active spots.

(2) The plate can be stained with iodine. This can be achieved rapidly, by shaking the plate in a bottle containing silica and a few crystals of iodine. The iodine will stain any compound that reacts with it and so is especially good for visualizing unsaturated compounds. Most spots show up within a few seconds, but the stain is not usually permanent.

(3) The plate can be treated with one of the reagents listed below and then heated to stain the spots. The reagent can be sprayed onto the plates, but this technique is quite hazardous and it is more effective for them to be dipped in the reagent. To do this, first let the TLC solvent evaporate, then holding the edge of the plate with tweezers, immerse the plate as completely as possible in the stain and remove it quickly. Rest the edge of the plate on a paper towel to absorb the excess stain before heating carefully on a hot plate or with a heat gun, until the spots show. This method is always permanent and so should be done last. When glass plates are used the spots can sometimes be seen more clearly from the glass side of the plate.

Stains	Use/comments
Vanillin	Good general reagent, gives a range of colours
PMA	Good general reagent, gives blue green spots
Anisaldehyde	Good general reagent, gives a range of colours
Ceric sulphate	Fairly general, gives a range of colours
DNP[#]	Mainly for aldehydes and ketones gives orange spots
Permanganate	Mainly for unsaturated compounds and alcohols, gives yellow spots

[#] usually do not require heating.

2.13.1.5 Recipes for visualization reagents

Vanillin: vanillin(15g) in ethanol(250mL) + conc. H_2SO_4 (2.5mL)

PMIA: phosphomlybdic acid(12g) in ethanol(250mL)

Anisaldehyde: anisaldehyde(15g) in ethanol(250mL) + conc. H_2SO_4 (2.5mL)

Ceric sulphate: 15% aqueous H_2SO_4 saturated with ceric sulphate

DNP: 2,4-dinitrophenolhydrazine (12g) + conc. H_2SO_4 (60mL) + water (80mL) + ethanol(2000mL)

Permanganllte: $KMnO_4$ (3g) + K_2CO_3 (20g) + 5% aqueous NaOH(5mL) + water (300mL)

2.13.1.6 Solvent systems and polarity

The distance which a compound travels up a TLC plate depends on two factors, its polarity and that of the solvent in the same solvent, the more polar the compound, the more tightly it is bound to the silica (or alumina) and the less it travels. There are some common trends, and with experience, it is often possible to predict whether a product will be less or more polar than the starting material. For example when a ketone, or ester is reduced, the resultant alcohol is almost always significantly more polar, and a clean transformation to a lower running spot will indicate a successful reaction. If the polarity of the solvent used for elution is increased the spots will move further up the plate and the distance between the centers of the spots usually increases, up to about half way up the plate. However, the spots also become more defuse, the further they travel, so an R_f value of about 0.4 is normally the optimum for analytical purposes.

The best TLC solvent system for a particular compound or mixture can only be determined by trial and error. However it is good practice to stick to a 'standard' solvent mixture, which can be used most of the time and you are familiar with. The most widely used solvent mixtures are based on a non-polar hydrocarbon, such as 40/60 petroleum ether or hexane, with a polar constituent added in a proportion which gives a suitable polarity. Probably the most popular 'universal' TLC system is petroleum ether-ethyl acetate, the polarity of which is easily adjusted by changing the proportions of the two solvents. If the compounds being analyzed will not travel in ethyl acetate mixtures, a more polar solvent such as ethanol is used as the additive. On the other hand, if the compounds travel too far a less polar additive such as petroleum ether is used.

The degree of separation between compounds will also vary according to the solvent used, so if compounds do not separate or give poor separation in one system, different systems should be tried. Where there are a number compounds in a mixture it may be best to use two or more different systems, for resolving the polar and non-polar components.

Common TLC solvents fall into one of three categories based on polarity, with smaller variations within each category.

Very polar solvent additives: methanol>ethanol>isopropanol
All much more polar than:
Moderately polar additives: acetonitrile>ethyl acetate>chloroform>dichloromethane>diethyl ether>toluene
All much more polar than:
Non-polar solvents: cyclohexane, petroleum-ether, hexane, pentane

Most solvent systems consist of one of the non-polar solvents together with a solvent from one of the other classes. However, for very polar compounds, one of the moderately polar solvents can be used as the less polar constituent. An example of this is chloroform-methanol mixtures, which are useful for highly hydroxylated compounds. The chlorinated hydrocarbons are also commonly used as single components.

R_f values

The R_f value of a compound depends upon the conditions under which the plate was run and is only accurate to about 20% therefore it is best to compare compounds on the same plate, and run a mixed spot. However it is useful to record the R_f value of a compound, remembering to quote the solvent system.

$$R_f = \frac{\text{Distance of centre of the spot from the baseline}}{\text{Distance of solvent front from the baseline}}$$

2.13.1.7 Multiple elutions

It is sometimes useful to elute a particular TLC plate several times and so improve the separation of closely running spots. In practice this is done by eluting the TLC plate as normal, removing it from the TLC tank and allowing the solvent to evaporate from TLC and then re-eluting the plate as before. Eluting a plate n-times is effectively the same as running a plate n-times its length.

Running acidic or basic compounds

Acidic and basic compounds often streak up the TLC plate. However, they will usually form distinct spots for acids a small amount of a carboxylic acid (e.g. acetic acid) is added to the solvent system, and for bases a small amount of amine (e.g. triethylamine) is added.

2.13.1.8 Running acid sensitive compounds

The silica on TLC plates is acidic in nature, and so compounds that are sensitive to acid may well decompose on TLC. There are several ways of getting around this problem, you can use alumina TLC plate (these suffer from the disadvantage that resolution is generally not as good, and The plates are basic in nature), or alternatively you can add a small amount of an amine (usually ammonia or triethylamine) to the solvent mixture to neutralize the acidic sites on the silica.

2.13.1.9 Checking for decomposition

Some organic compounds do decompose on silica to some extent, if you suspect that this is happening you can check by running a two-dimensional plate. This is done by cutting a square plate (ca. 5cm × 5cm) and spotting the compound in the bottom left-hand corner (ca. 0.5cm from the bottom) as shown in Fig. 2-42(a). The plate is then eluted as normal to give the spots in a line up the left hand side of the plate. The plate is then removed from the TLC tank and the solvent allowed to evaporate. It is then placed back in the tank with the line of spots along the bottom, and re-eluted [Fig. 2-42(b)].

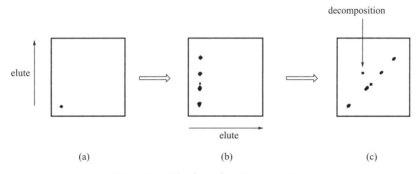

Fig. 2-42 Checking for decomposition

The result, if no decomposition is occurring will be the TLC running diagonally across the square plate, however if decomposition is occurring, the spot due to the unstable compound will show decomposition products off-diagonal [Fig. 2-42(c)].

It is very useful to carry out this test before any form of preparative chromatography, if you suspect that a compound may be labile on silica.

2.13.2 High Performance Liquid Chromatography (HPLC)

There is a broad range of HPLC techniques, and many different types of equipment. It is beyond the scope of this book to describe in great detail the methods for operating the equipment. This section will therefore focus on some of the ways that HPLC can be used to aid the synthetic organic chemist.

2.13.2.1 Description of HPLC

The general arrangement of an HPLC system is fairly simple, as shown in Fig. 2-43.
Solvent is pumped from a reservoir through a piston pump, which controls the flow rate. From the pump the solvent passes through a pulse damper, which removes some of the pulsing effect generated in the pump and also acts as a pressure regulator. In between the pulse damper and the column there is an injection valve which allows the sample to be introduced into the solvent stream. In the 'load' mode the solvent by-passes a sample loop, into which the sample is injected from a syringe. On switching to 'inject', the solvent stream is di-

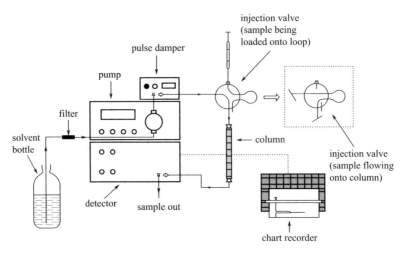

Fig. 2-43 The HPLC system

verted through the load loop, introducing a very accurately measured volume of the sample solution onto the column. Components are separated on the HPLC column in exactly the same way as they would be on a TLC plate, the less polar compounds running faster and coming through first. The effluent from the column passes into a detector (usually an ultraviolet or refractive index) which produces a signal on the chart recorder when a component is present.

The time at which the compound comes off the column is characteristic of that particular material, and is referred to as the retention time.

The area under any peak on the chart recorder is proportional to the quantity of that component and the method is therefore quantitative.

2.13.2.2 Uses of analytical HPLC

Finding a TLC system and running a sample can be done very quickly and for this reason TLC is the normal method of choice for routine reaction monitoring. However, there are occasions when it is worth spending the time to set up an HPLC system for reaction monitoring, especially if, as in many modern synthetic labs, you have a system close to hand. One reason to use HPLC is that the compounds in which you are interested do not separate very well on TLC. The other common reason is that you require a quantitative technique. This may be the case if you are trying to optimize a reaction to maximize the quantity of one product over another, and for this type of extended study it is well worth the time it takes to set up the system. For most synthetic purposes it is the relative, rather than the absolute proportions of substances which are important, and if that is the case a simple comparison of integrated peak areas may be all that is required. If accurate quantification is needed, then a calibration is required and this can be done using an internal standard (see below).

Another common use of HPLC is for identification of a compound by comparison with a known substance. Under a specific set of conditions (solvent, flow rate and quantity applied) any compound will have a specific retention time and this can therefore be used as a characteristic of the compound. However, just as a mixed spot should be always be run when comparing substances on TLC, so with HPLC a single enhanced peak should be observed when the comparison substance and the unknown are injected as a mixture. Again caution should be used, since a single peak is not absolute proof that compounds are the same.

Preparative HPLC is now becoming widely used in organic chemistry for separating compounds with, very similar polarity. Before committing all your material to a preparative column it is always best to run a small quantity of the sample on an analytical column, in order to work out the best conditions. Indeed, columns are produced in various sizes which are directly comparable with one another.

If you are monitoring a reaction by HPLC and you want to know the identity of one or more of the products, you can often separate a few milligrams from a few runs on the analytical column, which is enough to get a full range of spectral data. On simple HPLC systems this can be done manually by collecting the effluent from the column when the peak of interest is coming off, and repeating several times. On more sophisticated systems a fraction collector is often incorporated and in some cases injections can be made automatically, so that the system can be set up to collect a particular peak or peaks over a large number of runs.

Several methods are now available for coupling an HPLC system to a mass spectrometer, so that a mass spectrum is produced for each peak thus providing some structural information.

2.13.2.3 Quantitative analysis

For any type of detector each compound will have a different response, but for a particular compound, the area under its peak is directly proportional to the mass of material which produced the peak. If you are using a UV detector to analyze two compounds which have the same chromophore and extinction coefficients, then you can compare peak areas directly to determine the proportions of each compound.

Since the peak area is proportional to the mass of a compound, it is possible to take a known mass of a compound, make a standard solution and inject specific quantities, to work out the proportionality constant. However, a more accurate method of calibration is to use an internal standard. To do this follow the procedure below:

(1) Choose as a standard a readily-available stable compound, with a retention time away from the peaks of interest.

(2) Make up at least three mixtures containing known quantities of the standard and each of the compounds which are to be analysed.

(3) Run the mixtures and measure the areas of each peak

The mass of material under any peak, y, is:

$$M_y = k_y \cdot Area_y$$

now, comparing the area under the standard peak with that under the unknown:

$$\frac{M_y}{M_{(std)}} = \frac{k_y}{k_{(std)}} \times \frac{Area_y}{Area_{(std)}}$$

Using this equation we can work out $k_y/k_{(std)}$ which is a constant k_y, known as the correction factor for compound y. Using date from each of the runs, the average correction factor for each compound is calculated.

(4) Now if we want to calculate the quantity of a compound in a mixture, the mixture is spiked with a known quantity of the standard and the following equation is used:

$$M_y = \frac{k_y}{M_{(std)}} \times \frac{Area_y}{Area_{(std)}}$$

2.13.2.4 Peak shape

For good quantitative results from analytical HPLC (or GC, see below) you should aim to produce chromatographs with symmetrical peaks. Tailing of the peaks is usually caused by overloading and can thus be avoided by reducing the quantity of sample applied. If this does not solve the problem and the tail of a component is long and drawn out, there may be an incompatibility between the compound and the stationary phase, a problem which is less easy to rectify. HPLC can also be used for preparative separations.

2.13.3 Gas-liquid Chromatography (GC, GLC, VPC)

Gas-liquid chromatography is becoming increasingly popular for reaction monitoring and for analysis of reaction products. It can be used for the analysis of any compounds which are volatile below about 300℃ and thermally stable. It is not the intention of this section to give a detailed description of GC instrumentation, but simply to outline some of the uses of the technique for reaction monitoring and related work.

2.13.3.1 Description of gas chromatography

Gas chromatography is a very sensitive technique requiring only very small amounts of sample (10^{-6} g). A solution of about 1% is sufficient and a few microliter of this is injected into a heated injector block. A stream of carrier gas, usually helium, passes through the injector and sweeps the vapors produced onto the column, which is contained in an oven. The temperature of the oven can be accurately controlled and can either be kept constant or increased at a specified rate. Separation of the components in GC is not based on the principle of adsorption, as it is in liquid chromatography, but on partition. A GC column is rather like an extremely effective distillation column with the relative volatility of the components being the main factor which determines how quickly they travel through the column. The stationary phase of the column is a very high molecular weight, non-volatile oil, which has a very large surface area (see below for more detail about columns) and the gaseous components of the mixture are partitioned between the oil and the carrier gas at different rates. Thus the components are separated along the length of the column and emerge as discrete bands. The gas stream passing out of the column enters a flame ionization detector (FID) which produces an electric current when a compound is burned in this flame. The electric current is amplified to produce a peak on the chart recorder. These detectors are very sensitive and the response produced is proportional to the quantity of material being burned, thus the peak area is proportional to the quantity of sample. As with HPLC, the time taken for a particular substance to reach the detector is characteristic of that substance, and referred to as the retention time.

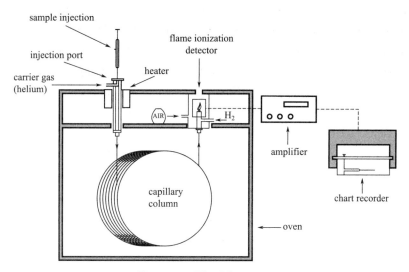

Fig. 2-44 The GC system

2.13.3.2 Types of GC column

A GC column must contain a liquid stationary phase with a large surface area, which must be supported, so that it stays in the column and the gas can pass through it. Packed columns are he traditional type, they are made from metal or glass coils, and have an internal diameter of about 2 to 4mm. The stationary liquid phase is coated on particles of solid with which the column is packed. There are a wide variety of stationary phases available ranging from Apiezon greases, which are very non-polar to polyethylene glycols which are very polar. With packed columns the choice of stationary phase is often critical for good separation and a good deal of experimentation is usually required before the best material is found.

The development of modern capillary columns has led to improved resolution and has also simplified the process of running GCs considerably. The columns are normally made from fused silica capillary with an inside diameter of between 0.2 and 0.5mm, and are polymer coated. They have no packing, but instead the liquid stationary phase is bonded to the inside wall of the capillary, and this allows gas to flow very easily. Because of this the columns can be made much longer than packed columns (between 12 and 100m) and they are typically ten times as efficient. Capillary columns give extremely high sensitivity and only a very small quantity of material is required. For this reason the injector normally incorporates a 'splitter', so that only a small portion of the sample injected actually enters the column.

The development of capillary columns has been largely responsible for the increased use of GC for monitoring organic reactions and for product analysis. The increased sensitivity of these columns is one important reason for this, but the fact that they are very simple to set up and operate is perhaps more significant The type of stationary phase is not so critical as with packed columns and the gas flow rate is essentially determined by the column, so the instrument can be operated successfully with minimal prior expertise. For most purposes relating to preparative organic chemistry it is sufficient to rely on just two types of column, one non-polar (such as a BP1) and one polar column (such as a BP20).

2.13.3.3　Uses of GC for reaction monitoring and product analysis

Capillary GC instruments are so simple to use that, provided there is one close by, monitoring a reaction by GC is almost as quick as running a TLC. It is common to turn to GC monitoring when TLC does not provide resolution between starting material and product or between one product and another. GC will usually separate components which co-run on TLC. We also find that some compounds, such as amines, which do not run very well on TLC, can be analyzed very easily by capillary GC.

GC also provides quantitative analysis and is widely used for determination of product ratios from diastereoselective reactions, down to about 200∶1. This makes it an ideal technique for optimization studies, where a large number of small-scale reactions are carried out under different conditions and product ratios are measured simply by syringing out a few microlitres from each and then injecting into the GC instrument earlier.

For absolute quantitative studies the GC instrument can be calibrated in exactly the same way as described for the HPLC instrument. Some people use this technique to measure theoretical yields for reactions, although it is always preferable that isolated yields are quoted.

The identity of a compound can often be determined by GC, if an unknown has the same retention time, and co-runs with a known compound when the two are injected as a mixture, but just as with TLC and HPLC co-running, caution should be exercised. A very powerful structure analysis technique is GC mass spectrometry (GC-MS) and when capillary GC is used this is a very simple and quick form of analysis. The mass spectrometer simply acts as the detector, but as well as providing a chromatograph, a mass spectrum of each individual component is obtained. Capillary GC-MS is a very sensitive technique and mass spectroscopic data can be obtained even on very minor components of a mixture.

One example of how GC and GC-MS can be useful is shown in Fig. 2-44. The first time this reaction was carried out, TLC analysis indicated that the starting material had been completely transformed to a single product, which ran as one spot in a number of solvent systems, but the product did not appear to be pure by ^{1}H NMR spectroscopy. When a GC-MS was run on the reaction product, two compounds (which ran together on TLC) where separated and easily distinguished as compounds A and C. also, 3% of isomeric compound B was identified and although this runs separately from A and C on TLC it had not been detected previously due to lack of sensitivity.

Having discovered the identity of the reaction products by GC-MS, product ratios for a series of reaction were obtained using simple GC and we were thus able to find optimum conditions for generation of compound A.

In this section we have not endeavored to give a comprehensive review of methods which can be used for reaction monitoring, but we have described the most universal and commonly used modern techniques. Various other monitoring methods can be devised for specific reactions. For instance ultraviolet (UV) spectroscopy can be used if one strong chromophore is being converted to another, but the disadvantage of spectroscopic methods is that they do not indicate how many products are being produced. NMR spectroscopy can also be employed.

2.14 Some Methods of Separation

2.14.1 Azeotropic and Extractive Distillation

Many mixtures are difficult to separate by distillation because their relative volatilities are close to 1.0. Some of these mixtures may be separated by the addition of another component that increase the relative volatilities of the original constituents. Fractionation after the addition of a component to increase relative volatilities is referred to a azeotropic distillation or as extractive distillation, depending upon the nature of the added component.

In azeotropic distillation the added component is relatively volatile, so that it appears in the distillate. In main cases it forms a low-boiling-point azeotrope with one of the original component. The low boiling azeotrope goes overhead in the distillate, and the remaining components go out at the bottoms. Of course, it is then necessary to separate the azeotropic agent from the original component by an appropriate separation process. Extractive distillation involves the addition of a relatively nonvolatile component. The 'solvent' added flows downward from stage to stage, increasing the relative volatilities. It finally flows out with the bottoms.

Azeotropic or extractive distillation is used to separate both binary and multicomponent mixtures. For simplicity, only binary mixtures are discussed here, but the principles and calculations are the same for multicomponent mixtures.

Azeotropic or extractive distillation is based on the interactions of the components of nonideal solutions. The azeotropic or extractive agent is chosen because it interacts more strongly with one of the components than it does with the other. Anazeotropic agent may be added to a nearly ideal binary mixture of low relative volatility (that is, the two components have nearly the same boiling point). If the structures of the original components are sufficiently different, the azeotropic agent may interact with one of them (for example, by hydrogen bonding, thereby increasing their relative volatility).

The original binary solution may be sufficiently nonideal to contain a binary azeotrope that blocks the production of pure components by simple distillation. Addition of an azeotropic agent may "break" the binary azeotrope and allow production of pure components, although it may require additional separation steps to recover the azeotropic agent. For example, ethyl alcohol and water form a binary azeotrope at 0.89 mole fraction alcohol, so that this is the richest alcohol solution that can be recovered by simple distillation. Addition of benzene as an azeotropic agent yields a distillate product consisting of a ternary azeotrope. The bottoms is the desired pure alcohol. Sufficient alcohol is lost in the ternary azeotrope to require its recovery.

On condensation, the ternary azeotrope in the distillate splits into two liquid phases. The benzene-rich phase is returned to the first distillation column, while the water-rich phase is sent to a second column. In this second column, the distillate is again the ternary azeotrope, which is condensed and sent to the phase splitter (together with the distillate from the first column). The bottoms from the second column is an aqueous solution of alcohol. It is sent to a third column that produces the water-alcohol azeotrope as the distillate and pure water as the bottoms. This distillate is mixed with the feed to the first column. Thus, the net effect of the process is to produce pure alcohol and pure water from the alcohol-water azeotrope.

Calculations for azeotropic distillation are nearly the same as those for simple multicomponent distillation. The addition of an azeotropic agent to a binary mixture produces a multicomponent mixture. Stage-to-stage calculations can be used to determine the separation possible with a given column or to determine the number of stages required for a given separation. Two differences arise. Equilibrium data are more complex because the K for a given component is a function not only of temperature and pressure, but also of the nature and concentration of other components present. Thus, complete nonideal equilibrium data must be available for the system of interset. A second difference arises from the fact that the azeotropic agent may be fed at more than one point in the column. Calculations must take into account these intermediate streams. In extractive distillation, the nonvolatile solvent is fed as an intermediate stream near the top of the column.

2.14.2 Supercritical Fluid Extraction

In supercritical-fluid extraction (SFE), a liquid or comminuted solid contacts a supercritical fluid at temperatures and pressures near the critical point and, thereby, permits the extraction of one or more components to be separated. The extraction is analogous to liquid extraction with the exception that the pressures and supercritical fluid's diffusivity are high and the fluid's density and viscosity are low, enabling rapid extraction and phase separation. Recovery of the extracted material is accomplished generally by lowering the pressure, which decreases, the fluid's density and markedly lowers its solubility, raising the fluid's temperature at constant pressure is another possibility. The cycle is completed by recompressing the gas to extraction conditions. The process is based on the significant changes in the solvent power of supercritical fluids, which occur with changes in temperature and pressure near the critical point.

The process is particularly well-suited for the separation of high boiling or heat-sensitive materials which are difficult to separate by distillation or liquid extraction. Low polarity materials tend to be preferentially extracted. The food industry has been one of the first to commercialize the process, because fluids, eg, CO_2, that pose no health hazards and leave no residue, can be used for extraction. In the FRG, caffeine is extracted commercially from coffee with supercritical CO_2; more recent developments are processes for the extraction of hops, spices, and tobacco. The pharmaceutical industry will likely find SFE useful. Supercritical-fluid extraction has also been examined extensively as a means of upgrading heavy hydrocarbons. Kerr-McGee Refining Corporation has developed the residuum oil supercritical extraction (ROSE) process in which supercritical pentane is used to deasphalt atmospheric and vacuum-distillation residua. The process is in limited commercial use. Kerr-McGee and the UK National Coal Board have also done considerable work on the SFE of coal and of coal liquids.

2.14.3 Liquid-Liquid Systems

The separation of the components of a liquid mixture by treatment with a solvent in which one or more of the desired components is preferentially soluble is known as liquid-liquid extraction-an operation which is used, for example, in the processing of coal tar liquids and in the production of fuels in the nuclear industry, and which has been applied extensively to the separation of hydrocarbons in the petroleum industry. In the operation it is essential that the liquid-mixture feed and solvent are at least partially if not completely immiscible and, in essence, three stages are involved:

(a) bringing the feed mixture and the solvent into intimate contact.

(b) separation of the resulting two phase.

(c) removal and recovery of the solvent from each phase.

It is possible to combine stages (a) and (b) into a single piece of equipment such as a column which is then operated continuously-such an operation is known as differential contacting. Liquid-liquid extraction is also carried out in stagewise equipment, the prime example being a mixer-settler unit in which the main features are the mixing of the two liquid phases by agitation, followed by settling in a separate vessel by gravity. This mixing of two liquids by agitation is of considerable importance.

For the great majority of reaction vessels, and for most operations involving liquid-liquid and to some extent liquid-solid mixing, the most commonly applied equipment involves a propeller or a turbine in a tank. It has been suggested that mixing of this type can be divided into three classes:

(a) Liquids, with or without solids, which remain free flowing when mixing is complete; e.g. water and salt, acid and sand, light or medium oils.

(b) Liquids, with or without solids, which are viscous but still pourable when mixing is complete; e.g. heavy oils, paints, resins, syrups.

(c) Liquids, with solids, which form stiff pastes; e.g. core sands and binders, oil-bound distempers, white lead and oil, putties.

2.14.4 Leaching

Leaching refers to the extraction of a soluble constituent from a solid by means of a solvent. The process may be employed either for the production of a concentrated solution of a valuable solid material, or in order to free an insoluble solid, such as a pigment, from a soluble material with which it is contaminated. The method used for the extraction will be determined by the proportion of soluble constituent present, its distribution throughout the solid, the nature of the solid and the particle size.

If the solute is uniformly dispersed in the solid, the material close to the surface will first be dissolved, leaving a porous structure in the solid residue. The solvent will then have to penetrate this outer layer before it can reach further solute, and the process will become progressively more difficult and the extraction rate will fall. If the solute forms a very high proportion of the solid, this porous structure may break down almost immediately to give a fine deposit of insoluble residue, and access of solvent to the solute will not be impeded. Generally the process can be considered in three parts: first the change of phase of the solute as it dissolves in the solvent, secondly its diffusion through the solvent in the pores of the solid to the outside of the particle, and thirdly the transfer of the solute from the solution in contact with the particles to the main bulk of the solution. Any one of these three may be responsible for limiting the extraction rate, though he first process usually occurs so rapidly that it has a negligible effect on the overall rate.

In some cases the soluble material is distributed in small isolated pockets in a material which is impermeable to the solvent such as gold dispersed in rock, for example. In such cases the material is crushed so that all the soluble material is exposed to the solvent. If the solid has a cellular structure, the extraction rate will generally be comparatively low because the cell walls provide an additional resistance. In the extraction of sugar from beet, the cell walls perform the important function of impeding the extraction of undesirable constituents of relatively high molecular weight and the beet should therefore be prepared in long strips so that a relatively small proportion of the cells ruptured. In the extraction of oil from seeds, the solute is itself liquid and may diffuse towards the solvent.

3 实　　验

3.1　烯烃的制备

实验1　环　己　烯

【目的与要求】
（1）掌握醇脱水制取烯烃的原理和方法。
（2）掌握微型蒸馏、微型分馏及微量液体干燥的方法。

【原理】
本实验采用环己醇在浓硫酸催化下脱水制备环己烯。

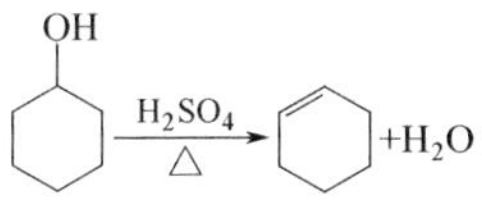

$$\text{环己醇} \xrightarrow[\Delta]{H_2SO_4} \text{环己烯} + H_2O$$

【仪器与药品】
（1）仪器：圆底烧瓶，电磁加热搅拌器，微型蒸馏头。
（2）药品：环己醇，浓硫酸，食盐，无水氯化钙，5%碳酸钠水溶液。

【步骤】
在25mL干燥的圆底烧瓶中加入7.5g环己醇、0.5mL浓硫酸[1]和沸石，充分摇振使之混合均匀。圆底烧瓶上安装微型分馏头，接上冷凝管及温度计，接收瓶浸在冷水中冷却。将圆底烧瓶在电磁加热搅拌器上用小火缓缓加热，控制加热速度缓慢地蒸出生成的环己烯和水，馏出温度不超过90℃[2]。馏出液为带水的浑浊液。至无液体蒸出且圆底烧瓶中只剩下很少量残液并出现阵阵白雾时，即可停止蒸馏。

馏出液用食盐饱和，然后加入1~2mL 5%的碳酸钠溶液中和微量的酸。将液体转入分液漏斗中，摇振后静置分层，分出有机相（哪一层？如何取出？），用1~2g无水氯化钙干燥至溶液清亮透明后滤入蒸馏瓶中[3]，加入沸石蒸馏，产量约为3g。

环己烯的沸点为82.98℃，折射率 n_D^{20} 为1.4465。其红外光谱和核磁共振谱见图3-1、图3-2。

【附注】
［1］本实验也可用3mL 85%的磷酸代替浓硫酸作脱水剂，其余步骤相同。
［2］用油浴进行加热使蒸馏瓶受热均匀。由于反应中环己烯与水形成共沸物（沸点为70.8℃，含水10%），环己醇与环己烯形成共沸物（沸点为64.9℃，含环己醇30.5%），环己醇与水形成共沸物（沸点为97.8℃，含水80%）。因此，在加热时温度不可过高，蒸馏速度不宜太快，以减少未作用的环己醇蒸出。

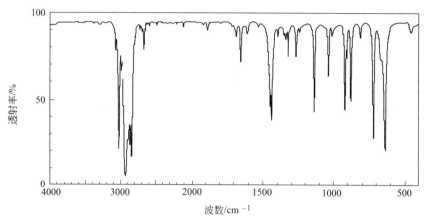

图 3-1　环己烯的红外光谱图

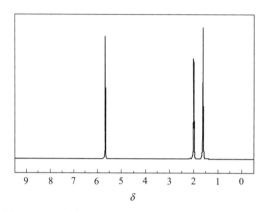

图 3-2　环己烯的核磁共振谱图（溶剂：四氯化碳）

[3] 水层应尽可能分离完全，否则将增加无水氯化钙的用量，使产物更多地被干燥剂吸附而导致收率降低。这里用无水氯化钙干燥较适宜，它还可除去少量环己醇（生成醇与氯化钙的配合物）。

思 考 题

(1) 在粗制环己烯中，加入食盐使水层饱和的目的何在？
(2) 写出无水氯化钙吸水后的化学变化方程式，为什么蒸馏前一定要将它过滤掉？
(3) 写出下列醇与浓硫酸进行脱水的产物。

　　3-甲基-1-丁醇　　　3-甲基-2-丁醇　　　3,3-二甲基-2-丁醇

Experiment 2　Synthesis of 2-Methyl-2-butene

【Experimental Objectives】

(1) Study atmospheric distillation operation.
(2) Master the lab preparation method of olefin.

【Experimental Principles】

The dehydration of the appropriate alcohols is one of the effective method to obtain the alkene in the lab. Based on the hygroscopicity of concentrated sulfuric acid, it can be used as the dehydrating reagent in the dehydration reaction. Herein, amyl alcohols can be dehydrated to obtain 2-methyl-2-butene[1] by the concentrated sulfuric acid.

$$\underset{\underset{OH}{|}}{\overset{\overset{CH_3}{|}}{H_3C-C-CH_2CH_3}} \xrightarrow{H_2SO_4} \overset{\overset{CH_3}{|}}{H_3C-C=\underset{\underset{H}{|}}{C}-CH_3}$$

【Apparatus and Reagents】

(1) Apparatus: Round bottom flask, Vigreux fractionating column, straight condenser, bent adapter, thermometer, thermowell, take-off adapter, receiving flasks, water bath.

(2) Reagents: Concentrated sulfuric acid, water, t-amyl alcohol.

【Experimental Procedure】

Cautiously add 5mL of concentrated sulphuric acid slowly and with constant stirring to 10mL of water contained in a small beaker.

Cool the dilute acid, transfer it to the flask[2], add 8mL of t-amyl alcohol and a few fragments of porous porcelain. Assemble the apparatus completely, making sure that all the corks are secure, and arrange for the flask to be heated on a water bath. Heat gently and when distillation commences, regulate the temperature of the bath so that the temperature on the thermometer does not exceed 40—41℃ (1—2 drops per second). Stop the distillation when the temperature can no longer be maintained below 41℃. The product is practically pure, but contains a little water.

Transfer the distillate to a small conical flask and dry it over 0.3—0.5g of anhydrous magnesium sulphate or calcium chloride. The yield is 3—3.5g.

【Notes】

[1] Pure 2-methyl-2-butene boils at 38.5℃.

[2] The flask must be well stoppered owing to the volatility of the hydrocarbon.

3.2 卤代烃的制备

实验3 溴 乙 烷

【目的与要求】

(1) 学习用醇与HX通过亲核反应制备卤代烷的方法。

(2) 掌握低沸点有机物的微型蒸馏操作。

【原理】

反应式：

$$NaBr + H_2SO_4 \rightleftharpoons NaHSO_4 + HBr$$

$$CH_3CH_2OH + H^+ \rightleftharpoons CH_3CH_2\overset{+}{O}H_2$$

$$CH_3CH_2O^+H_2 + Br^- \longrightarrow CH_3CH_2Br + H_2O$$

副反应：

$$2NaBr + 3H_2SO_4 == Br_2 + SO_2 + 2H_2O + 2NaHSO_4$$

$$CH_3CH_2OH \begin{array}{c} \xrightarrow{H^+} CH_3CH_2OCH_2CH_3 \\ \xrightarrow{H^+} CH_2=CH_2 \end{array}$$

【仪器与药品】

(1) 仪器：10mL 圆底烧瓶，直形冷凝管，分液漏斗，蒸馏头。

(2) 药品：无水乙醇，溴化钠，浓硫酸，无水氯化钙。

【步骤】

在 10mL 圆底烧瓶中，放入 2.0mL H_2O，2.0mL 无水乙醇，慢慢加入 4.0mL 浓硫酸，混合均匀，并冷却到室温。加入 3.0g 研细的溴化钠，充分振荡，按图 2-1(c) 安装仪器，接收瓶内外均放冰水。缓慢加热，溴化钠溶解，蒸馏直至无油滴滴出。用毛细管小心将馏出液中的有机层吸入试管中，振荡下加入浓硫酸 0.2mL，吸去浓硫酸层，加入 1 小块无水氯化钙干燥。过滤出的粗产物进行蒸馏，收集 38～40℃ 的馏分，产量约为 1.6g。

思 考 题

本实验产率不高的原因何在？

实验 4 溴 丁 烷

【目的与要求】

(1) 学习用醇与 HX 通过亲核反应制备卤代烷的方法。

(2) 练习吸收有害气体的回流加热操作。

(3) 巩固微型蒸馏操作，学习微量液体的洗涤、干燥等操作。

【原理】

反应式：

$$NaBr + H_2SO_4 == NaHSO_4 + HBr$$

$$CH_3CH_2CH_2CH_2OH + H^+ \rightleftharpoons CH_3CH_2CH_2CH_2\overset{+}{O}H_2$$

$$CH_3CH_2CH_2CH_2O^+H_2 + Br^- \longrightarrow CH_3CH_2CH_2CH_2Br + H_2O$$

副反应：

$$2NaBr + 3H_2SO_4 == Br_2 + SO_2 + 2H_2O + 2NaHSO_4$$

$$CH_3CH_2CH_2CH_2OH \begin{array}{c} \xrightarrow{H^+} CH_3CH_2CH_2CH_2OCH_2CH_2CH_2CH_3 \\ \xrightarrow{H^+} CH_3CH_2CH=CH_2 \end{array}$$

【仪器与药品】

(1) 仪器：10mL 圆底烧瓶，球形冷凝管，直形冷凝管，二通活塞，导气管，分液漏斗，蒸馏头，玻璃漏斗。

(2) 药品：正丁醇，溴化钠，浓硫酸，饱和碳酸氢钠，无水氯化钙。

【步骤】

在 10mL 圆底烧瓶中,放入 2.0mL H_2O,慢慢加入 2.4mL 浓硫酸,混合均匀,并冷却到室温。加入 1.5mL 正丁醇,混合后,再加入 2g 研细的溴化钠,充分振荡,按图 1-2(c) 装置仪器。加热回流 30min。然后冷却,取下气体吸收装置和球形冷凝管,装上微型蒸馏头和球形冷凝管,蒸馏出粗产物溴丁烷[1]。

粗产物溴丁烷分别用 2mL 水、1mL 浓硫酸[2]、2mL 饱和碳酸氢钠溶液洗涤,再用 1 小块无水氯化钙干燥。过滤出的粗产物进行蒸馏,收集 99～103℃ 的馏分,产量约为 1.2g。

【附注】

[1] 粗产物溴丁烷是否蒸完,可以根据以下几点来判断:①馏出液是否由浑浊变澄清;②蒸馏烧瓶上的油层是否消失;③馏出液滴入水中,是否有油珠。

[2] 浓硫酸可除去原料正丁醇及副产物丁醚等。

思 考 题

(1) 本实验如何减少副反应的发生?

(2) 蒸馏出粗产物溴丁烷后的残余物主要是什么,为何要趁热倒入烧杯中?

实验 5 α,β-二溴苯乙烷

【目的与要求】

(1) 学习用烯烃与卤素 X_2 发生加成反应制备卤代烷的方法。

(2) 掌握重结晶等操作。

【原理】

苯乙烯与卤素很容易发生亲电加成反应,反应式如下:

$$C_6H_5CH=CH_2 + Br_2 \longrightarrow C_6H_5CHBrCH_2Br$$

α,β-二溴苯乙烷的相对分子质量为 264,为针形结晶,m.p.:72～73℃,b.p.:139～141℃/2kPa,不溶于水,溶于乙醇、乙醚、苯、醋酸等。

【仪器与药品】

(1) 仪器:10mL 圆底烧瓶,球形冷凝管,砂芯漏斗。

(2) 药品:苯乙烯,溴素,四氯化碳,甲醇。

【步骤】

装有冷凝管的 10mL 圆底烧瓶中加入 0.5mL 苯乙烯,滴加溴的四氯化碳溶液 3mL(浓度为 0.3g/mL)[1],并摇动反应瓶,此时有二溴化产物的结晶生成。当反应完全后,加入 5mL 甲醇[2],加热回流,如果结晶不能完全溶解,补加少量的甲醇,再回流,重复进行,直至结晶完全溶解。放置冷却,抽滤,干燥,测定熔点。

【附注】

[1] 溴易挥发,具有强烈的刺激性和腐蚀性,反应需在通风橱内进行。

[2] 甲醇的加入是为了对产品进行重结晶,因此甲醇的量以控制在回流温度下产品刚能溶解为宜。

思 考 题

下述化合物与溴发生亲电加成反应的活性次序如何?

Experiment 6 Synthesis of *t*-Butyl Chloride

【Experimental Objectives】

(1) Learn the preparation of halogenated alkanes by alcohols and haloid acid *via* nucleophilic reaction.

(2) Master the microscale distillation processes for organic compounds with low boiling points.

【Experimental Principles】

Tertiary alcohols can be converted to alkyl chlorides very readily. In this experiment, *t*-butyl alcohol can be converted to *t*-butyl chloride by shaking the alcohol in a separatory funnel with concentrated HCl, then the product can be purified by a simple distillation.

When mix the reactants, the mixture is initially a homogeneous solution because *t*-butyl alcohol is miscible with water (Concentrated HCl is a 12mol/L aqueous solution; therefore, it is mostly water). *t*-Butyl alcohol is more soluble than 1-butanol (solubility= 8.3g/100mL) because its branched hydrocarbon chain is more compact and thus less hydrophobic than a four-carbon continuous chain. Immediately after the reactants are mixed, a second phase is formed. This second phase is the water-insoluble *t*-butyl chloride. The difference in solubilities between the alcohol and the halide is attributed to the hydrogen bonding between *t*-butyl alcohol and water and the lack of hydrogen bonding between *t*-butyl chloride and water.

$$(CH_3)_3C-OH + HCl \longrightarrow (CH_3)_3C-Cl + H_2O$$

The reaction of a tertiary alcohol and a hydrohalic acid proceeds by way of a carbocation (S_N1 reaction). The carbocation combines with a halide ion (the nucleophile) to yield the product alkyl halide. However, the carbocation can also lose a proton to a molecule of water or alcohol to yield an alkene. Therefore, when *t*-butyl alcohol is treated with conc. HCl, some methylpropene (isobutylene) is also formed. Because elimination reactions are favored by high temperatures, chilling the HCl before mixing it with the alcohol helps suppress the formation of the alkene.

Step 1: protonation

$$(CH_3)_3C-\ddot{O}H + H^+ \overset{fast}{\rightleftharpoons} (CH_3)_3C-\overset{+}{O}H_2$$

t-butyl alcohol → protonated *t*-butyl alcohol (an oxonium ion)

Steric hindrance prevents backside nucleophilic attack

Step 2: ionization

$$(CH_3)_3C\overset{+}{-}\overset{..}{O}H_2 \underset{}{\overset{slow}{\rightleftharpoons}} (CH_3)_3C^+ + H_2\overset{..}{\underset{..}{O}}$$

protonated *t*-butyl alcohol　　　　　　　　　*t*-butyl-cation　　water

Step 3: attack of nucleophile

$$(CH_3)_3C^+ + \overset{..}{\underset{..}{Cl}}:^- \overset{fast}{\rightleftharpoons} (CH_3)_3C-\overset{..}{\underset{..}{Cl}}:$$

2-chloro-2-methylpropane
(*t*-butyl chloride)

Methylpropene is a gas under the reaction conditions; therefore, it evaporates from the reaction mixture as it is formed. No special work-up procedure is necessary to remove this by-product from the desired product. However, when the reaction mixture is being shaken in the separatory funnel, the funnel must be vented frequently to release the gas. Furthermore, when the separatory funnel is sitting in an iron ring, its stopper must be removed to allow the methylpropene to escape.

After the reaction is complete, it is necessary to separate the *t*-butyl chloride from the excess acid. *t*-Butyl chloride is less dense than water; consequently, it forms the upper layer in the separatory funnel. The bulk of excess acid and water is removed simply by draining the lower layer. The residual acid in the separatory funnel is removed by a water wash, followed by a wash with aqueous $NaHCO_3$. The *t*-butyl chloride is then washed with water to remove any residual salts. Finally, the wet *t*-butyl chloride is dried with anhydrous $CaCl_2$ and distilled.

The drying should be done promptly because *t*-butyl chloride reacts slowly with water at room temperature to yield *t*-butyl alcohol and HCl in a solvolysis reaction. If the *t*-butyl chloride is wet when it is distilled, noticeable amounts of HCl will be evolved.

t-Butyl chloride is extremely volatile. Because evaporation decreases the yield, any container of *t*-butyl chloride should be kept tightly stoppered between laboratory periods or during drying. To minimize evaporation during the distillation, the receiving flask should be chilled in an ice bath.

【Apparatus and Reagents】

(1) Apparatus: Heating mantle, electromagnetic stirrer, round bottom flask, separatory funnel, distilling head, thermometer, spherical condenser, straight condenser, take-off adapter, conical flask.

(2) Reagents: *t*-Butanol, concentrated hydrochloric acid, sodium bicarbonate, anhydrous calcium chloride.

【Experimental Procedure】

Place *t*-butanol (5mL) and concentrated hydrochloric acid (15mL) in a round-bottom flask with a reflux condenser and an electromagnetic stirrer, stir for 3min. The reaction can also be carried out by shaking the alcohol in a separatory funnel with concentrated HCl solution. Transfer the mixture into a separatory funnel, then stand until two layers of the liquid separate obviously. Remove the lower acid layer, wash the upper organic layer sequentially with 5% sodium bicarbonate solution and water to neutral, dry with anhydrous calcium chloride.

Transfer the crude product to a dry distilling flask. Add a small amount of zeolite, distill and collect the boiling portion at 51—52℃ in a weighed conical flask.

【Thinking Questions】

(1) Can we prepare the *n*-butyl chloride with the same conditions by 1-butanol and concentrated hydrochloric acid?

(2) Which compound exhibits higher boiling point, *n*-butyl chloride or *t*-butyl chloride?

3.3 醇 的 制 备

实验 7 苯 甲 醇

【目的与要求】

(1) 掌握苯甲醇的制备方法。

(2) 掌握相转移催化反应技术。

【原理】

$$C_6H_5CH_2Cl \xrightarrow[H_2O]{K_2CO_3} C_6H_5CH_2OH + KCl + CO_2$$

【仪器与药品】

(1) 仪器：电磁加热搅拌器，三口烧瓶，球形冷凝管，滴液漏斗。

(2) 药品：苄氯，碳酸钾，四乙基溴化铵，无水硫酸镁，乙醚。

【步骤】

在装有电磁加热搅拌器的25mL三口烧瓶里加入碳酸钾水溶液（0.8g碳酸钾溶于10mL水中）及0.2mL 50%四乙基溴化铵水溶液[1]，加入沸石。装上回流冷凝管和滴液漏斗，在滴液漏斗中装1mL苄氯。在回流状态下将苯氯甲烷滴入三口烧瓶中，滴加完毕后，在搅拌下回流反应1h。

反应完毕后，冷却到30~40℃[2]。把反应液移入分液漏斗中，分出油层。碱液用乙醚萃取3次，每次用2mL乙醚。合并乙醚液和粗苯甲醇。用无水硫酸镁或碳酸钾干燥。将干燥透明的苯甲醇乙醚溶液先在热水浴上蒸出乙醚，然后加热分馏。收集200~208℃的馏分，产量约为0.5g。

纯苯甲醇为无色透明液体，b. p. 为205.4℃，d_4^{20} 为1.0419，n_D^{20} 为1.5396。其红外光谱见图3-3。

【附注】

[1] 也可用其他相转移催化剂，如三乙基苄基溴化铵。

[2] 如果温度过低，碱易析出给分离带来困难。

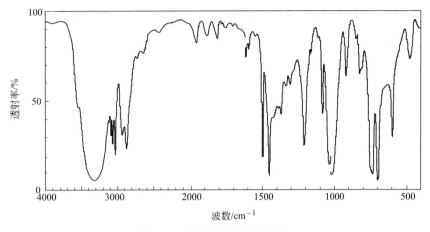

图 3-3　苯甲醇的红外光谱图

思 考 题

(1) 实验室还有哪些方法可用来制备苯甲醇？
(2) 本实验为什么采用碳酸钾作为苯氯甲烷的碱性水解剂？

实验 8　二 苯 甲 醇

【目的与要求】

(1) 掌握酮还原制备仲醇的原理和方法。
(2) 进一步掌握过滤和重结晶等基本操作。

【原理】

$$C_6H_5COC_6H_5 \xrightarrow{Zn+NaOH} C_6H_5CH(OH)C_6H_5$$

【仪器与药品】

(1) 仪器：25mL 锥形瓶，抽滤瓶、砂芯漏斗。
(2) 药品：二苯甲酮，锌粉，氢氧化钠，乙醇，浓盐酸，石油醚（60～90℃）。

【步骤】

在装有冷凝管的 25mL 的锥形瓶中，依次加入 0.75g 氢氧化钠，0.75g 二苯甲酮，0.75g 锌粉和 10mL 95%的乙醇。充分摇振使反应微微放热约 20min 后，在80℃的水浴上加热 5min 使其反应完全。反应毕，用真空抽滤，固体用少量乙醇洗涤。滤液倒入 40mL 冷水中，摇荡均匀后用浓盐酸小心酸化至 pH 值为 5～6[1]，再真空抽滤析出的固体。粗产物干燥后，用石油醚[2]重结晶就得到二苯甲醇的针状晶体。测量其熔点。m.p.：68～69℃。

【附注】

[1] 酸化时溶液酸性不宜太强，否则固体难以析出。
[2] 也可用己烷作为重结晶溶剂。

思 考 题

(1) 二苯甲醇还可以用其他什么方法合成？
(2) 除用锌粉还原以外，还可以用什么还原剂？

Experiment 9 Synthesis of Triphenylmethanol

【Experimental Objectives】

(1) Learn the principle and operation method of preparing triphenylmethanol by Grignard reaction.

(2) Master the waterless operation technology.

【Experimental Principles】

Triphenylmethanol can be synthesized readily by the Grignard reaction. Phenylmagnesium bromide is prepared by direct reaction of bromobenzene with metallic magnesium in the presence of anhydrous diethyl ether (or tetrahydrofuran).

$$C_6H_5Br + Mg + (C_2H_5)_2O \longrightarrow C_6H_5MgBr \cdot (C_2H_5)_2O$$

Usually a crystal of iodine is added to aid in starting the reaction, which must be carried out with carefully purified reagents and under anhydrous conditions. A small amount of biphenyl, $C_6H_5\text{-}C_6H_5$, is formed through coupling of the aryl groups (Wurtz-Fittig reaction).

$$C_6H_5MgBr + C_6H_5Br \longrightarrow C_6H_5-C_6H_5 + MgBr_2$$

Reaction of phenylmagnesium bromide with ethyl benzoate gives triphenyl methanol as the major product.

$$C_6H_5MgBr + C_6H_5COOC_2H_5 \longrightarrow (C_6H_5)_2CO + C_2H_5OMgBr$$
$$C_6H_5COC_6H_5 + C_6H_5MgBr \longrightarrow (C_6H_5)_3COMgBr$$

Acidic hydrolysis of the reaction mixture affords an organic layer (ethereal solution) containing biphenyl, ethyl benzoate, triphenylmethanol, benzophenone and benzene, the latter coming from reaction of water with phenylmagnesium bromide.

$$C_6H_5MgBr \xrightarrow[H^+]{H_2O} C_6H_6$$

The principal organic products present are the desired triphenylmethanol and the unwanted biphenyl. The crude product may be purified by recrystallization.

【Apparatus and Reagents】

(1) Apparatus: Electromagnetic heating stirrer, three-necked flask, spherical condenser, dropping funnel.

(2) Reagents: Magnesium, bromobenzene, anhydrous diethyl ether, diethyl ether, 5%-10% sulfuric acid, strong sodium bisulfite, crystal of iodine.

【Experimental Procedure】

Equip a 100mL dry three-necked, round-bottomed flask with a mechanical stirrer, a reflux condenser and a dropping funnel. Attach calcium chloride drying tube to the top of the condenser. Introduce directly into the flask 1.2g (0.05mol) of magnesium turning, a mixture of 1mL (1.5g) of bromobenzene, 4mL of anhydrous diethyl ether and a small crystal of iodine. If a reaction does not start at once, warm the flask gently in a bath of warm water. After the reaction has started, add 35mL of anhydrous diethyl ether.

Place in the addition funnel 4.5mL (6.75g) of bromobenzene (a total of 0.0525mol) and allow it to flow drop by drop into the previously activated reaction mixture at such a rate that the ether refluxes without external heating.

After all of the halide has been added, reflux the mixture gently for 30 min on a steam bath. Do not heat the material so vigorously that ether vapors traverse the condenser. The reaction is complete when the magnesium has dissolved some dark particles of impurities will remain undissolved. Remove the heating bath and proceed without delay to the next step.

Cool the reaction flask containing the Grignard reagent to 15—20℃ and place in the addition funnel a solution of 4mL (4g, 0.025mol) of pure ethyl benzoate in about 12mL of anhydrous ether. Allow the ethyl benzoate solution to flow slowly into the Grignard reagent, with continuous stirring, and cool the flask from time to time to control the reaction. The bromomagnesium derivative of the alcohol separates as a white precipitate. After all of the ethyl benzoate has been added, allow the mixture to stand at room temperature for 30min or longer.

Pour the contents of the flask as completely as possible into a mixture of about 25g of ice, 50mL of water, and 2—3mL of concentrated sulfuric acid, contained in a 250mL flask. Add 2—3mL of strong sodium bisulfite solution to remove any free iodine. Shake the mixture thoroughly to complete the decomposition of the magnesium derivative and rinse the reaction flask with the acid mixture to remove material that adheres to the wall of the flask. Add about 40mL of ordinary diethyl ether to aid in extracting the product completely. Separate the ether layer, wash it with two 10mL portions of 5%—10% sulfuric acid and once with saturated salt (NaCl) solution. Finally, wash the ether layer with aqueous sodium bicarbonate solution and once more with saturated salt solution.

Dry the ether solution over anhydrous magnesium sulfate and transfer it to a 250mL round-bottom flask arranged for distillation. Distill off the ether as completely as possible, using a bath (Avoid fire hazards!). The residual crude product contains the impurities biphenyl and unreacted bromobenzene and ethyl benzoate, along with triphenylmethanol.

Stir the residue with petroleum ether or ligroin (about smug of residue) and collect the solid on a suction filter this process removes most of the impurities. Complete the purification by recrystallization from 2-propanol (about 7mL/g), and collect the product on a suction filter. The yield is 3.5—4g (Literature: m. p. 162—164℃).

【Notes】

[1] In Grignard reactions, it is essential that the reagents be free from ethanol and water, and the apparatus perfectly clean and dry.

[2] The clean, unoxidized surfaces of magnesium must be used.

[3] It is essential that the ether be of the anhydrous grade.

[4] It is evidenced by disappearance of the iodine color, appearance of turbidity, and spontaneous boiling.

[5] For the success of the experiment it is essential that the reaction begin before the main portions of the ether and bromobenzene are added.

【Thinking Questions】

(1) Ethanol is often present in the technical grade of diethyl ether. If this grade rather than anhydrous were used, what effect, if any, would the ethanol have on the formation of the Grignard reagent? Explain?

(2) Why is it unwise to allow the solution of the Grignard reagent to remain exposed to air for an unnecessary period of time even if it is protected from moisture by drying tubes?

(3) Why is it unwise to begin addition of be solution of ethyl benzoate to the Grignard reagent before the latter has cooled to room temperature?

(4) What is the solid that forms during the addition of the ester to the Grignard reagent?

3.4 醚 的 制 备

实验 10 乙 醚

【目的与要求】

(1) 学习并掌握低沸点物质向高温反应液中滴加的方法。

(2) 学习由醇脱水制备醚的方法。

【原理】

醚是常用的优良溶剂,有的反应(如格氏反应)必须在酸中进行。因此,在有机合成中,醚有着广泛的应用。醚可由相应的醇通过分子间脱水制备,所用的催化剂可以是硫酸或氧化铝。反应式为:

$$R-OH \xrightleftharpoons[\triangle]{催化剂} R-O-R + H_2O$$

本实验以硫酸为催化剂,用乙醇脱水制备乙醚。醇与硫酸的作用,随温度不同而生成不同的产物。

反应式:

$$CH_3CH_2OH + H_2SO_4 \xrightleftharpoons{室温} [CH_3CH_2OH_2]^+ [HSO_4]^-$$

$$CH_3CH_2OH + H_2SO_4 \xrightleftharpoons{100℃} CH_3CH_2OSO_3H + H_2O$$

$$2CH_3CH_2OH \xrightleftharpoons{H_2SO_4, 140℃} CH_3CH_2OCH_2CH_3 + H_2O$$

$$CH_3CH_2OH \xrightarrow{H_2SO_4, >160℃} CH_2=CH_2 + H_2O$$

【仪器与药品】

(1) 仪器:25mL三口圆底烧瓶,球形冷凝管,温度计,直形接引管,多功能梨形漏斗,H形分馏头,锥形瓶,分液漏斗,10mL圆底烧瓶,电热套。

(2) 药品:95%乙醇,浓硫酸,5%氢氧化钠溶液,饱和氯化钠溶液,饱和氯化钙溶液,无水氯化钙。

【步骤】

在干燥的25mL三口圆底烧瓶中,加入2.5mL 95%乙醇。在冷水浴冷却下,边摇动边慢慢滴加 2.5mL 浓硫酸，加毕，充分摇动使其混合均匀，并加入 1~2 粒沸石。在三口烧瓶的一个侧口安装直形接引管和多功能梨形漏斗，另一侧口安装200℃温度计。直形接引管末端及温度计水银球均应浸入液面以下，距瓶底 0.5~1cm 处。三口瓶的中口装有 H 形分馏头，将其上口加塞封住，另一口装有球形冷凝管，用一锥形瓶接收液体，锥形瓶用冷水冷却，H 形分馏头的支管接橡皮管，将尾气导入下水道。装置如图 1-5 所示。

在多功能梨形漏斗中，放入5mL 95%乙醇，用电热套加热三口瓶，使反应温度较快地升至140℃[1]，这时，开始向漏斗中慢慢滴入 95%乙醇，控制滴入速度与馏出速度大致相等[2]（每秒 1~2 滴），并控制液温于 135~150℃之间。待 95%乙醇加完后（0.5~1h），继续加热 10min，当温度上升到160℃时，撤去热源，停止反应。

将锥形瓶中的馏出液转移到分液漏斗中，依次用2mL 5%氢氧化钠溶液、2mL 饱和氯化钠溶液洗涤，最后每次用 1.5mL 饱和氯化钙溶液洗涤 2 次。分去水层，将乙醚层倒入干燥锥形瓶中，用少许无水氯化钙干燥。

将干燥后的液体滤入 10mL 圆底烧瓶中，加入沸石，安装常压蒸馏装置，用电热套缓缓加热，收集 33~38℃馏分。称重。产量为 1.7~1.8g，产率为 35%~37%。

纯乙醚：b.p. 为34.51℃，n_D^{20} 为 1.3526，d_4^{20} 为 0.7138。其红外光谱见图 3-4。

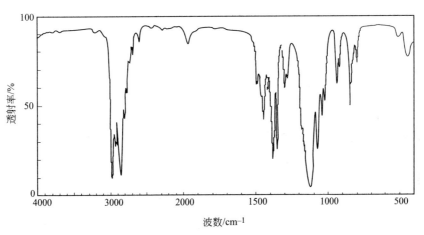

图 3-4 乙醚的红外光谱图

【结果与讨论】

（1）反应现象

收集粗产品温度 34~50℃左右。反应液变成黑色液体。洗涤时醚层在上。

（2）实验结果

借助直形接引管使乙醇顺利滴入反应混合物液面下，有利于提高乙醚的产量。但由于反应物加量少，反应温度控制不够稳定，以致很快升到160℃，生成了部分乙烯，降低了乙醚的产率，因此微型实验产品产率比常量实验偏低。本实验的关键是控制反应温度。

【附注】

[1] 控制加热，使温度较快地升到100℃以上，然后稍降电压使液温不超过150℃。

[2] 在140℃时就有乙醚馏出，此时再滴入乙醇，就可继续与硫酸氢乙酯作用生成乙醚。所以滴入乙醇的速度应与乙醚馏出的速度相等。若滴加过快，不仅乙醇未及时作用就被蒸出，且使反应液的温度骤降，减少乙醚的生成。

思 考 题

(1) 用浓硫酸催化法由乙醇制备乙醚,为什么乙醇-浓硫酸混合液加热到乙醇沸点以上时,乙醇不会被蒸出?如在滴加乙醇时去掉直形接引管,将产生什么后果?

(2) 在用饱和氯化钙溶液洗涤之前,为何先用饱和氯化钠溶液洗?

(3) 在用硫酸处理乙醇和正丙醇的混合物时,可以得到三种醚的混合物,它们是什么?而用硫酸处理叔丁醇和乙醇的混合物时,却得到产率很好的一种醚,它可能是什么醚?如何解释?

实验11 正 丁 醚

【目的与要求】

(1) 掌握醇脱水制醚的方法。
(2) 掌握分水器的操作。

【原理】

主反应:

$$2CH_3CH_2CH_2CH_2OH \underset{134\sim135℃}{\overset{H_2SO_4}{\rightleftharpoons}} (CH_3CH_2CH_2CH_2)_2O + H_2O$$

副反应:

$$CH_3CH_2CH_2CH_2OH \underset{>135℃}{\overset{H_2SO_4}{\rightleftharpoons}} C_4H_8 + H_2O$$

【仪器与药品】

(1) 仪器:二口烧瓶,微型蒸馏头,球形冷凝管,分水器,分液漏斗,电热套。
(2) 药品:正丁醇,硫酸,饱和食盐水,无水氯化钙。

【步骤】

在25mL二口烧瓶中加入8mL正丁醇,边摇动边慢慢加入1mL浓硫酸,充分摇动使之混合均匀[1],并加入沸石。在烧瓶上口装上分水器,分水器上端连接球形冷凝管。分水器中可以加入少量的水。将烧瓶放在电热套中缓慢加热,保持回流反应。随着反应的进行,分水器中的水层不断增加[2],反应液的温度也逐渐上升。如果分水器中的水层超过了支管而流回烧瓶时,可打开旋塞放掉一部分水。当烧瓶中反应液温度达到150℃左右时,反应完毕,停止加热[3]。

待反应物冷却后,将装置改装成蒸馏装置,重新加入沸石,进行蒸馏至无馏出液为止。将馏出液倒入分液漏斗中,分去水层。粗产物用2mL冷的50%硫酸洗涤2次[4],再用水洗涤2次,最后用1~2g无水氯化钙干燥。将干燥后的粗产物进行蒸馏(**注意:不要把氯化钙倒进去**),收集140~144℃馏分,产量约为1.5g。

纯正丁醚为无色液体,b. p. 为142.4℃。

【附注】

[1] 充分摇动以防止硫酸局部过浓,否则加热时会使反应溶液变黑。

[2] 本实验利用恒沸混合物蒸馏方法将反应生成的水不断从反应物中除去。正丁醇、正丁醚和水可以生成以下几种恒沸混合物,见表3-1。

表 3-1 正丁醇、正丁醚和水的恒沸混合物指标

恒沸混合物		沸点/℃	组分的质量分数/%		
			正丁醚	正丁醇	水
二元	正丁醇-水	93.0		55.5	45.5
	正丁醚-水	94.1	66.6		33.4
	正丁醇-正丁醚	117.6	17.5	82.5	
三元	正丁醇-正丁醚-水	90.6	35.5	34.6	29.9

这些含水的恒沸物冷凝后，在分水器中分层。上层主要是正丁醇和正丁醚，下层主要是水。利用分水器就可以使上层有机物流回到反应器中。

[3] 反应开始回流时，因为有恒沸物的存在，温度不可能马上达到150℃。但随着水被蒸出，温度逐渐升高，最后可达到150℃以上，此时应立即停止加热。如果温度升得太高，反应溶液会炭化变黑，并有大量副产物丁烯生成。

[4] 用50%硫酸可以除去粗制正丁醚中的正丁醇，因正丁醇能溶于浓硫酸中，而正丁醚则很少溶解。

思 考 题

(1) 精制时，各步洗涤的目的是什么？

(2) 如果最后蒸馏前得出产品中含有丁醇，能否用分馏的方法将它除去？这样做好不好？

Experiment 12　　Synthesis of 4-Nitroanisol

【Experimental Objectives】

(1) Learn the principle of etherification.

(2) Master the operations of filtration.

【Experimental Principles】

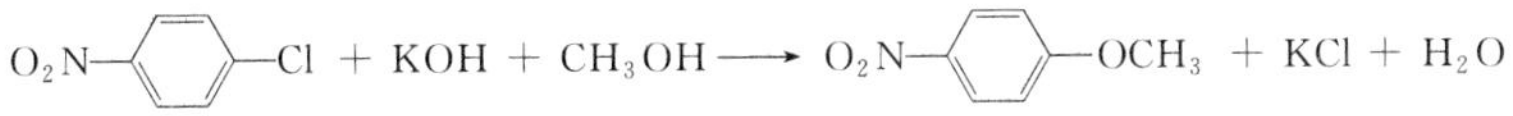

$$O_2N\text{-}C_6H_4\text{-}Cl + KOH + CH_3OH \longrightarrow O_2N\text{-}C_6H_4\text{-}OCH_3 + KCl + H_2O$$

【Apparatus and Reagents】

(1) Apparatus: Three-necked round bottom flask, condenser pipe, thermometer, mechanical stirrer, water bath.

(2) Reagents: 4-Nitrochlorobenzene, KOH, CH_3OH.

【Experimental Procedure】

1.96g (0.0125mol) 4-nitrochlorobenzene, 2.1g (0.0375mol) KOH, and 5mL CH_3OH are added into a 25mL flask. Mechanical stirrer is used to mix them. Then the mixture is heated by water bath to keep the temperature at 65℃ for 3h. Under this reaction process, some white solid (KCl) precipitate in the flask. When the reaction finish, the mixture is poured into ice water, and collect the white precipitation. Then wash the solid until to the neutral pH. After drying the solid, 1.75g white solid (m.p. 52—53℃) is collected.

【Notes】

The reaction temperature should not too high or heat up too fast.

3.5 醛酮的制备

实验13 环 己 酮

【目的与要求】

(1) 掌握铬酸氧化制备环己酮的原理和方法。
(2) 进一步掌握微型萃取的操作。

【原理】

反应式:

$$3 \bigcirc\!\!-\!\!OH + Na_2Cr_2O_7 + 5H_2SO_4 \longrightarrow 3 \bigcirc\!\!=\!\!O + Cr_2(SO_4)_3 + 2NaHSO_4 + 7H_2O$$

【仪器与药品】

(1) 仪器：圆底烧瓶，冷凝管，分液漏斗，微型分馏头，烧杯。
(2) 药品：环己醇，重铬酸钠，浓硫酸，氯化钠，无水硫酸镁。

【步骤】

在圆底烧瓶内加入 9mL 冰水，边摇动烧瓶边慢慢加入 1.5mL 浓硫酸，再小心加入 1.5mL 环己醇。将溶液冷却至室温。

在烧杯内将 1.5g 重铬酸钠溶于 1.5mL 水中。将此溶液冷却到室温，并慢慢的分批加到环己醇的硫酸溶液中。不断地摇动烧瓶以使反应物充分混合。第一批重铬酸钠溶液加入后，反应物温度自行上升，反应体系由橙红色变成墨绿色[1]。待反应物温度升到55℃时，用冷水适当冷却以控制反应温度不超过60℃。待反应物的橙红色完全消失后，方可加下批重铬酸钠溶液。重铬酸钠溶液全部加完后继续摇动烧瓶，直至反应温度开始下降。加入 0.2～0.5mL 的甲醇或 0.2g 的草酸。反应毕，在反应物内加入 7.5mL 水及沸石，把环己酮以水蒸气蒸馏蒸出来[2]，直到馏出液澄清为止。在馏出液中加入氯化钠饱和水[3]。将液体移入分液漏斗中分出有机层，水层用 3mL 乙醚萃取 2 次。合并有机层，用无水硫酸镁干燥。蒸除乙醚后蒸馏，收集 151～156℃的馏分，产量约为 0.6g。

纯环己酮为无色液体，沸点为 155.7℃，n_D^{20} 为 1.4507，环己酮的红外光谱见图 3-5。

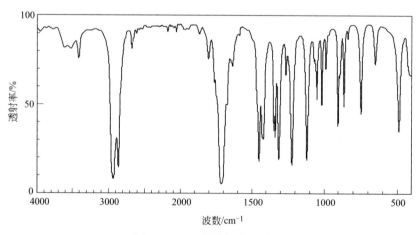

图 3-5 环己酮的红外光谱图

【附注】

[1] 反应开始后，重铬酸盐（橙红色）转变为低价铬盐（墨绿色）。

[2] 环己酮和水形成共沸混合物，沸点为95℃，含环己酮38.4%。

[3] 在水中的溶解度为2.4g/100g水。加入氯化钠是为了降低环己酮的溶解度，并增加水的密度，有利于环己酮的分层。

思 考 题

（1）在加重铬酸钠溶液过程中，为什么要待反应物的橙红色完全消失后方能加下一批重铬酸钠？在整个氧化反应过程中，为什么要控制温度在一定的范围？

（2）氧化反应结束后，为什么要往反应物中加入甲醇或草酸？

（3）从反应混合物中分离环己酮除了在采用水蒸气蒸馏外，还可以采用何种方法？

实验 14 苯 乙 酮

【目的与要求】

（1）掌握傅-克酰基化制备芳酮的原理和方法。

（2）掌握回流、萃取、干燥等操作。

【原理】

$$\underset{}{\text{C}_6\text{H}_6} + (CH_3CO)_2O \xrightarrow{AlCl_3} C_6H_5COCH_3 + CH_3COOH$$

【仪器与药品】

（1）仪器：三口圆底烧瓶，直形冷凝管，短颈漏斗，弯干燥管，多功能梨形恒压漏斗，茄形瓶，H形分馏头，分液漏斗，电磁加热搅拌器。

（2）药品：无水苯，无水三氯化铝，醋酐，浓盐酸，苯，5%氢氧化钠溶液，无水硫酸镁。

【步骤】

按图3-6安装好仪器。称取3.0g研细的无水三氯化铝[1]放入三口圆底烧瓶，再加入4.5mL（4.0g，51mmol）无水苯。在电磁搅拌下，由滴液漏斗慢慢加入1.5mL（1.3g，17mmol）无水苯和1.07g（10mmol）醋酐的混合物。加完后，使其在水浴上加热到无氯化氢气体产生为止，约需15min。

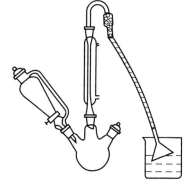

图3-6 苯乙酮制备装置

将三口圆底烧瓶浸入冷水浴上，搅拌反应液，用滴液漏斗慢慢滴入9mL浓盐酸和3mL冰水的混合液[2]。在滴加过程中先出现白色沉淀，继续滴加，沉淀溶解，液体分层。将其转入分液漏斗中，分出油层。水层用苯萃取3次，每次2mL，合并有机层，再依次用2mL 5%氢氧化钠溶液和2mL水洗涤1次，有机层用适量无水硫酸镁干燥。将干燥后的液体直接抽滤到二口瓶中，先在电热套上蒸馏出苯，再减压蒸馏，收集78℃/1.333kPa（10mmHg）、98℃/3.325kPa（25mmHg）的馏分，产量约为0.6g。苯乙酮的红外光谱见图 3-7。

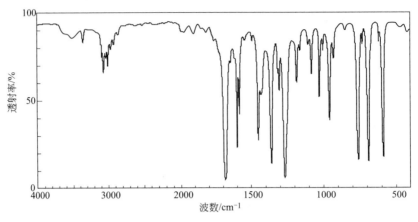

图 3-7 苯乙酮的红外光谱图

【附注】

[1] 无水三氯化铝极易接触空气吸水分解，所以称量研细时必须迅速。整个反应仪器也必须干燥。

[2] 苯乙酮/三氯化铝络合物分解释放出大量的热，必须慢慢滴加浓盐酸。

思 考 题

(1) 水和湿气对本实验有何影响？仪器和装置应注意哪些问题？

(2) 无水三氯化铝在本实验中的作用是什么？

实验 15　茴香基丙酮

【目的与要求】

(1) 了解香料合成的一些基本常识，熟悉茴香基丙酮的生产工艺。

(2) 掌握无水试剂的使用方法。

(3) 熟练掌握分馏及减压蒸馏的基本操作。

【原理】

茴香基丙酮是一种无色或淡黄色油状液体，m.p.：9～10℃，b.p.：277℃（106℃/22mmHg），极微量溶于水，溶于乙醇。具有强烈的鲜花和水果香气，低浓度时具有樱桃-覆盆子的香味。其化学名称为 4-(对甲氧基苯基)-2-丁酮。天然的茴香基丙酮存在于沉香木（Aquilaria agallocha Roxb）的提取液及水解所得精油中。它是医药多巴酚丁胺的中间体，现是一种比较高档的香料，用来配制各种水果型香料，用在软饮料、冷饮、糖果、食品中。

茴香基丙酮的合成方法有许多种，但大多数工艺条件要求较高。本实验的合成路线如下所示。

$$\text{C}_6\text{H}_5\text{-OCH}_3 \xrightarrow[\text{AlCl}_3]{\text{CH}_2=\text{CH-COCH}_3} \text{H}_3\text{CO-C}_6\text{H}_4\text{-CH}_2\text{CH}_2\text{COCH}_3$$

【仪器与药品】

(1) 仪器：三口圆底烧瓶，冷凝管，电磁加热搅拌器，滴液漏斗，吸滤瓶。

(2) 药品：苯甲醚，甲基乙烯酮，环己烷，无水 $AlCl_3$，食盐，无水氯化钙，盐酸。

【步骤】

在装有电磁加热搅拌器、滴液漏斗和附有无水氯化钙干燥管的回流冷凝管的 25mL 三口圆底烧瓶[1]中，加入 8mL 环己烷和 7g 无水 $AlCl_3$ 混合搅拌，然后加入 9g 苯甲醚。用冰水冷却，将温度控制在 15℃ 以下，滴加 3g 甲基乙烯酮，搅拌 1h 后[2]，再将反应物小心倒入 10mL 冰水中，加少量的盐酸，使析出的沉淀溶解，分液，有机层用 10mL 食盐水洗涤 2 次，蒸馏回收环己烷和苯甲醚，再减压蒸馏，收集 154～156℃/1.87kPa 的馏分，测定折射率（文献值 $n_D^{20}=1.5198$）。

【附注】

[1] 无水氯化钙对湿气非常敏感，会与水迅速反应，因此，所用的玻璃仪器均需充分干燥。

[2] 反应中有氯化氢放出，所以本实验应该在通风橱中进行。

思 考 题

本实验的主要副反应是哪一个？

Experiment 16 Synthesis of Benzophenone

【Experimental Objectives】

(1) Learn the application of Friedel-Crafts in the synthesis and the reaction of kaolinite to ketone.

(2) Learn the principle of benzophenone synthesis.

【Experimental Principles】

Benzophenone can be used as a photo initiator in UV-curing applications such as inks, imaging, and clear coatings in the printing industry. Benzophenone prevents ultraviolet (UV) light from damaging scents and colors in products such as perfumes and soaps.

Benzophenone can also be added to plastic packaging as a UV blocker to prevent photo-degradation of the packaging polymers or its contents. Its use allows manufacturers to package the product in clear glass or plastic (such as a PETE water bottle). Without it, opaque or dark packaging would be required.

In biological applications, benzophenones have been used extensively as photophysical probes to identify and map peptide-protein interactions.

Benzophenone is produced by the copper-catalyzed oxidation of diphenylmethane with air.

A laboratory route involves the Friedel-Crafts acylation of benzene with benzoyl chloride in the presence of a Lewis acid (e.g. aluminium chloride) catalyst. It can also be prepared by reaction of benzene with carbon tetrachloride followed by hydrolysis of the resulting diphenyldichloromethane.

This reaction is catalyzed by anhydrous aluminum chloride. All the apparatus and reagents in the reaction should be dry and anhydrous because aluminum chloride will invalidate by decomposition if it encounters water or moisture. So it is desirable to introduce a drying tube between the condenser and the trap. The evolution of hydrogen chloride take place in this experiment, it is needed to fit with a gas absorption trap.

$$2 \, C_6H_6 + CCl_4 \xrightarrow{\text{anhydrous AlCl}_3} (C_6H_5)_2CCl_2 + 2HCl$$

$$(C_6H_5)_2CCl_2 + H_2O \longrightarrow (C_6H_5)_2C=O + 2HCl$$

【Apparatus and Reagents】

(1) Apparatus: Electromagnetic heating stirrer, spherical condenser, oil bath, 100℃ thermometer, three-necked round bottom flask, separatory funnel.

(2) Reagents: Anhydrous benzene, carbon tetrachloride, anhydrous aluminum chloride, magnesium sulfate.

【Experimental Procedure】

In a 50mL three-necked round bottom flask fitted with a good mechanical stirrer[1], a separatory funnel, a thermometer, and a reflux condenser connected with a trap for an absorbing the hydrogen chloride evolved, are placed anhydrous aluminum chloride (0.9g) and dry carbon tetrachloride (2mL). The flask is surrounded by an ice bath[2]. The stirred is started and when the temperature of carbon tetrachloride has dropped to 10—15℃, a mixture of anhydrous benzene (1.2mL) and carbon tetrachloride (1mL) is added at such a scale that the temperature is kept between 5 and 10℃. The reaction begins immediately as is indicated by the evolution of hydrogen chloride and a rising temperature. This addition requires about 15min. The stirring is continued for another one hour after the benzene-carbon tetrachloride solution has been added, while the temperature is held at about 10℃. The reaction mixture is still kept in an ice bath, and 1mL water is slowly added with stirring. After addition, the equipment is changed to distillation. The mixture is then first heated on a steam bath to remove most of the excess of carbon tetrachloride, and then the mixture is distilled with heating to carry over the remaining carbon tetrachloride, and to hydrolysis the benzophenone dichloride to benzophenone. The upper benzophenone layer is then separated from the aqueous layer and the latter is extracted with benzene (10mL). Combine the benzene solution and the benzophenone and dry them over anhydrous magnesium sulfate. The benzene solution and benzophenone are transferred to a flask for distillation. The benzene is removed under ordinary pressure and the benzophenone is distilled under reduced pressure collecting the distillate at 187—190℃/15mmHg[3]. The product is solidified and the yield is about 0.9—1g, m.p. 47—48℃. (Lit. m.p. 48.1℃).

【Notes】

[1] The mechanical stirrer should be very efficient, as otherwise the aluminum chloride tends to cake on the sides of the flask. This makes cooling very difficult and thus in-

crease the time necessary for the addition of the benzene-carbon tetrachloride mixture.

[2] It is necessary to allow the reaction to start before packing in an ice-salt mixture. If the temperature is too low (below 10℃) the reaction does not start. After the reaction has started, the cooling should be as efficient as possible so that the mixture of benzene and carbon tetrachloride may be added in the minimum amount of time. If the temperature drops below 5℃ the reaction is too slow. If the temperature goes above 10℃ there is increasing formation of tarry matter and lowering of the yield.

[3] There is considerable tendency for the benzophenone to foam over during the early part of the distillation under reduced pressure and care must be taken to prevent this.

【Thinking Questions】

(1) How is this experiment influenced by water and moisture? Which notes should we pay attention to in the apparatus and operation?

(2) Why should carbon tetrachloride rather than benzene be excessive? What is the consequence if benzene is excessive?

(3) What are the different quantities of aluminum chloride in the alkylation and acylation? Why?

Experiment 17　Synthesis of Diphenylethanedione (Benzil)

【Experimental Objectives】

(1) Understand how to choose the oxidizer in the synthetic process.

(2) Master the operation of reflux reaction and recrystallization.

【Experimental Principles】

Benzil (systematically known as 1,2-diphenylethane-1,2-dione) is the organic compound with the formula $(C_6H_5CO)_2$, generally abbreviated $(PhCO)_2$. This yellow solid is one of the most common diketones. Its main use is as a photoinitiator in polymer chemistry.

Most benzil is used in the free-radical curing of polymer networks. Ultraviolet radiation decomposes benzil, generating free-radical species within the material, promoting the formation of cross-links. Benzil is a potent inhibitor of human carboxylesterases, enzymes involved in the hydrolysis of carboxylesters and many clinically used drugs.

Benzil is a relatively poor photoinitiator, and is seldom used. It absorbs at the 260nm wavelength. It undergoes photobleaching, which allows the curing light to reach deeper layers of the material on longer exposure.

Benzil is prepared from benzoin, for example with copper(Ⅱ) acetate. Other suitable oxidizing agents such as nitric acid (HNO_3) are used routinely.

【Apparatus and Reagents】

(1) Apparatus: Round bottom flask, water bath, beaker, Buchner funnel, glass rod.

(2) Reagents: Benzoin, concentrated nitric acid, cold water.

【Experimental Procedure】

Place crude benzoin (2g) and concentrated nitric acid (10mL) in a 25mL round bottom flask. Heat on a boiling water bath (in the fume cupboard) with occasional shaking until the evolution of oxides of nitrogen has ceased (about 1-5h). Pour the reaction mixture into 30—40mL of cold water contained in a beaker, stir well until the oil crystallises completely as a yellow solid. Filter the crude benzil at the pump, and wash it thoroughly with water to remove the nitric acid. Recrystallise from alcohol or methylated spirit (about 2—6mL per gram). The yield of pure benzil, m.p. 94—96℃, is 1.9g.

3.6 羧酸的制备

实验18 苯甲酸

【目的与要求】

掌握高锰酸钾氧化甲苯制备苯甲酸的方法。

【原理】

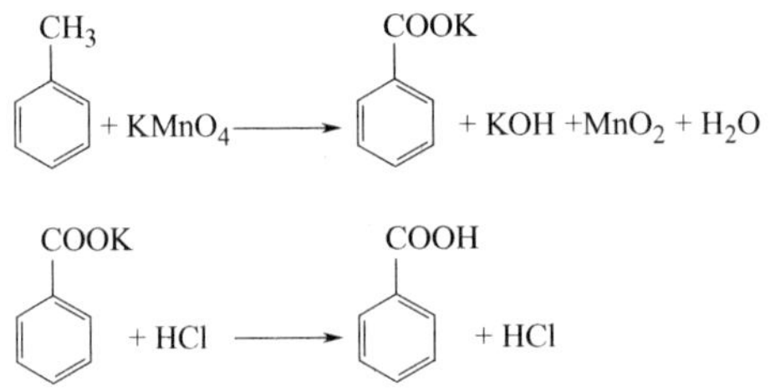

$$\text{C}_6\text{H}_5\text{CH}_3 + \text{KMnO}_4 \longrightarrow \text{C}_6\text{H}_5\text{COOK} + \text{KOH} + \text{MnO}_2 + \text{H}_2\text{O}$$

$$\text{C}_6\text{H}_5\text{COOK} + \text{HCl} \longrightarrow \text{C}_6\text{H}_5\text{COOH} + \text{HCl}$$

【仪器与药品】

(1) 仪器：圆底烧瓶，冷凝管，电磁加热搅拌器，漏斗，吸滤瓶。

(2) 药品：甲苯，高锰酸钾，浓盐酸。

【步骤】

在圆底烧瓶中加入1mL甲苯和20mL水，上口装回流冷凝管，通过电磁加热搅拌器搅拌加热回流后从冷凝管上口分批加入3g高锰酸钾[1]。黏附在冷凝管内壁的高锰酸钾最后用9mL水冲洗入瓶内。继续回流反应直到甲苯层几乎近于消失、回流液不再出现油珠（约需4～5h）。

将反应混合物趁热减压过滤[2]，用少量热水洗涤二氧化锰滤渣。合并滤液和洗涤液，放在冰水浴中冷却，然后用浓盐酸酸化至刚果红试纸变蓝，晶体析出。抽滤，沉淀用少量冷水洗涤后抽干、烘干得粗产品。粗品在水中重结晶即得纯品，产量约为0.5g。

纯苯甲酸为无色针状晶体，熔点为122.4℃。

【附注】

[1] 高锰酸钾每次加量不宜太多，否则反应异常剧烈。必须待反应平缓后再加下一批。

[2] 滤液如果呈紫色，可加入少量亚硫酸氢钠使紫色褪去，重新抽滤。

思 考 题

（1）氧化反应中影响苯甲酸产量的重要因素有哪些？
（2）加亚硫酸氢钠的目的是什么？

实验19　对硝基苯甲酸

【目的与要求】
掌握铬酸氧化芳烃制备芳基甲酸的方法。

【原理】

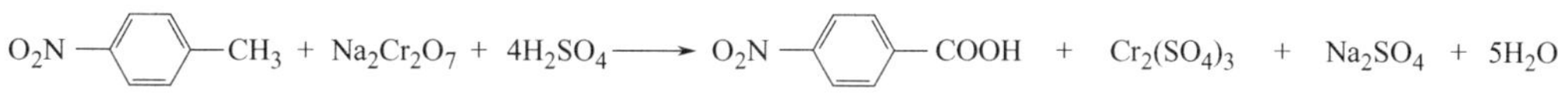

$$O_2N-C_6H_4-CH_3 + Na_2Cr_2O_7 + 4H_2SO_4 \longrightarrow O_2N-C_6H_4-COOH + Cr_2(SO_4)_3 + Na_2SO_4 + 5H_2O$$

【仪器与药品】
（1）仪器：圆底烧瓶，回流冷凝管，电磁加热搅拌器，滴液漏斗，吸滤瓶，烧杯。
（2）药品：对硝基甲苯，重铬酸钠，浓硫酸，5%氢氧化钠溶液，15%硫酸，活性炭。

【步骤】
将1g对硝基甲苯、3g重铬酸钠和4mL水加入25mL圆底烧瓶内。圆底烧瓶上口装回流冷凝管，用电磁加热搅拌器边搅拌边从冷凝管上口分批加入5mL浓硫酸。随着浓硫酸的加入，反应物的温度迅速上升，对硝基甲苯熔化，反应混合物的颜色逐渐变深。严格控制加入浓硫酸的速度（最好用滴液漏斗！），保持反应混合物的温度低于沸腾温度，必要时可用冷水冷却[1]。

硫酸加完后，加热回流反应0.5h。反应毕，冷却烧瓶，在搅拌下将反应混合物缓慢地倒入15mL冷水中，即有沉淀析出。减压过滤，尽量压碎粗产物，并用8mL水分2次洗涤滤饼。粗制的对硝基苯甲酸为深黄色固体。为了除去夹杂的铬盐，将粗产物移入烧杯中，加入12mL 5%氢氧化钠溶液，温热（不超过60℃）溶解后抽滤[2]，滤液中加入0.5g活性炭煮沸后热过滤。在搅拌下将滤液慢慢地倒入10mL 15%硫酸中，此时浅黄色沉淀立即析出[3]。减压过滤，滤饼用少量水洗涤几次，抽去水分，烘干。产物用乙醇-水重结晶精制，产品为浅黄色针状结晶，产量约为0.9g。熔点为241～242℃。

【附注】
[1] 如果温度过高，对硝基甲苯挥发易凝结在冷凝管壁上。
[2] 溶液中的不溶物为氢氧化铬和未反应的对硝基甲苯。
[3] 硫酸不能反加至滤液中，否则生成的沉淀会包含一些钠盐而影响产品的纯度。中和时溶液应呈强酸性。

思 考 题

（1）加入硫酸的速度为什么不能快？
（2）粗产品可能含有哪些杂质？如何除去？
（3）脱色后为什么不能将硫酸反加至滤液中？

实验 20 己 二 酸

【目的与要求】
(1) 了解环己醇氧化制备己二酸的原理与方法。
(2) 掌握气体吸收、滴液和重结晶等操作。

【原理】

$$\text{环己醇} \xrightarrow{HNO_3} \text{环己酮} \xrightarrow{HNO_3} HOOC(CH_2)_4COOH$$

【仪器与药品】
(1) 仪器：三口圆底烧瓶，回流冷凝管，电磁加热搅拌器，滴液漏斗，吸滤瓶，温度计。
(2) 药品：硝酸，环己醇，钒酸铵。

【步骤】
本实验必须在通风橱内进行。做实验时必须严格地遵照规定的反应条件进行。

在三口圆底烧瓶中插入温度计（其水银球要尽量接近瓶底）和回流冷凝管，回流冷凝管上端接气体吸收装置，用碱液吸收反应产生的氧化氮气体[1]。在烧瓶中加5mL水、5mL硝酸和1小粒钒酸铵。搅拌将溶液混合均匀，在水浴上加热到50℃。用滴管滴加2滴环己醇。反应立即开始，温度随即上升并有红棕色气体放出。小心地逐渐滴加2.1mL环己醇（最好用滴液漏斗！），控制温度在85~90℃，必要时用冷水冷却[2]。待环己醇全部加入后维持混合物在85~90℃下反应2~3min，至几乎无红棕色气体放出为止。

将反应物倾入冷水冷却，抽滤析出的晶体。烧瓶中剩余的晶体用滤液洗出。用3mL冰水洗涤己二酸粗品2次[3]，抽滤，产量约为1.4g。粗品用水重结晶后测定熔点。

纯己二酸是无色单斜晶体，熔点为153℃。己二酸的红外光谱见图3-8。

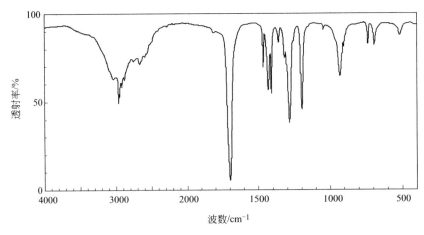

图 3-8 己二酸的红外光谱图

【附注】
[1] 因产生的氧化氮气体有毒，实验最好在通风橱内进行。废气不可逸散在实验室内。

〔2〕此反应为强放热反应,不可大量加入。否则反应过于剧烈容易引起爆炸。

〔3〕己二酸为二元酸,水溶性较大,洗涤时需用冰水,若浓缩母液可回收少量产物。其溶解度数据见表 3-2。

表 3-2 己二酸的溶解度

温度/℃	15	34	50	70	87	100
溶解度/(g/100g 水)	1.44	3.08	8.46	34.1	94.8	100

思 考 题

(1) 实验时为什么必须控制环己醇的滴加速度和反应物的温度?

(2) 本实验为什么必须在通风橱内进行?

实验 21 香豆素-3-羧酸

【目的与要求】

(1) 了解 Knoevenagel 缩合反应原理和香豆素-3-羧酸的制备方法。

(2) 掌握回流、重结晶等操作。

【原理】

本实验以水杨醛和丙二酸二乙酯在六氢吡啶存在下发生 Knoevenagel 缩合反应制得香豆素-3-羧酸酯,然后在碱性条件下水解制得目标产物。反应式:

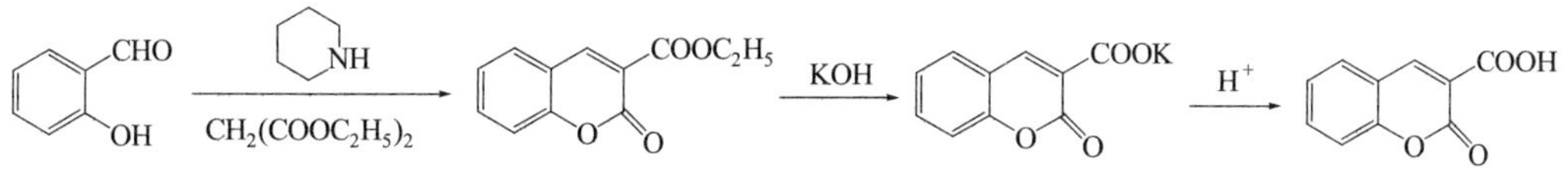

【仪器与药品】

(1) 仪器:磁力搅拌机,圆底烧瓶,电热套,球形冷凝管,滴液漏斗,分液漏斗,布氏漏斗,吸滤瓶。

(2) 药品:水杨醛,丙二酸二乙酯,六氢哌啶,冰醋酸,无水乙醇,浓盐酸,氢氧化钾,无水氯化钙。

【步骤】

在 25mL 圆底烧瓶中依次加入 1mL 水杨醛、1.2mL 丙二酸二乙酯、5mL 无水乙醇和 0.1mL 六氢吡啶及一滴冰醋酸,在无水条件下搅拌回流 1.5h,待反应物稍冷却后移走干燥管,从冷凝管顶端加入 6mL 冷水,待结晶析出后抽滤并用 1mL 被冰水冷却过的 50% 乙醇洗两次,得到粗品 3.3g,粗品可用 25% 乙醇重结晶,干燥后得到香豆素-3-甲酸乙酯,熔点 93℃。

在圆底烧瓶中加入 0.4g 香豆素-3-甲酸乙酯、0.3g 氢氧化钾、2mL 乙醇和 1mL 水,加热回流 15min。趁热将反应产物倒入 10mL 浓盐酸和 5mL 水的混合物中,立即有白色晶体析出,冰浴冷却后过滤,用少量冰水洗涤,干燥后的粗品约 0.8g,可用水(2~3mL)重结晶。

计算产率,测定产物熔点,熔点为 190℃左右。

Experiment 22 Synthesis of Cinnamic Acid

【Experimental Objectives】

(1) Understand the preparation principle and method of cinnamic acid.

(2) Master the operation of steam distillation.

(3) Master the purification methods for solid organic compounds: decolorization and recrystallization.

【Experimental Principles】

According to Perkin reaction, heat the mixture of aromatic aldehyde and acetic anhydride with the existence of the relative metallic salt of a carboxylic acid, and get α,β-unsaturated acid. For example:

$$PhCHO + (CH_3CO)_2O \xrightarrow{CH_3COOK} PhCH=CHCOOH + CH_3COOH$$

However, in this experiment, the method put forward by Kalnin is used, potassium carbonate instead of potassium acetate is adopted, the reaction time is shorter and the percentage yield is better.

【Apparatus and Reagents】

(1) Apparatus: Three-necked round bottom flask, beaker, glass rod, cylinder, thermometer (200℃), straight condenser, electromagnetic heating stirrer, spherical condenser, surface dish, filter paper, Buchner funnel, suction flask, conical flask.

(2) Reagents: Benzaldehyde, acetic anhydride, anhydrous potassium carbonate, 10% sodium hydroxide solution, hydrochloric acid, 20% alcohol, Congo red test paper.

【Experimental Procedure】

In a 200mL round bottom flask, place 0.75mL (0.78g, 7.5mmol) fresh distilled benzaldehyde, 2mL (2.16g, 21mmol[1]) fresh distilled acetic anhydride and 1.85g anhydrous potassium carbonate powder[2]. Heat the mixture on a wire gauze to reflux for 0.5h. As there is some carbon dioxide escaping, bubbles can be found in the beginning.

Pour 5mL water after cooling the contents of the flask for some time. The solid formed in the flask is to be made into pieces as much as possible. Then distillate off benzaldehyde left with steam distillation. Cool the flask again pour 5mL 10% sodium hydroxide solution to make sure that all the cinnamic acid be changed into sodium salt and dissolve in the water. Use vacuum filtration and pour the filtrate into a 25mL beaker, then cool it to room temperature. Add successive small portions of hydrochloric acid, while stirring the mixture, until the Congo red pH paper turns blue. Cool it and collect the crystals with suction. Wash the product with a small amount of cold water, and spread it on a clean watch glass to dry. The yield is about 0.75g (65%—70%). Then the rude product should be recrystalated with 20% alcohol. The melting point of pure cinnamic acid is 133℃.

【Notes】

[1] It is essential that benzaldehyde and acetic anhydride should be fresh distillate.

[2] Anhydrous potassium carbonate should be dried at 120℃ for 1h.

【Thinking Questions】

(1) Why is it essential that benzaldehyde and acetic anhydride should be fresh distillate?

(2) What product will be obtained when benzaldehyde reacts with anhydrous propionic anhydride in the presence of potassium propionate?

(3) Why may potassium acetate be instead by potassium carbonate?

3.7 羧酸衍生物

实验 23 邻苯二甲酸二丁酯

【目的与要求】

(1) 学习邻苯二甲酸二丁酯的制备原理和合成方法。

(2) 掌握分水器的使用方法。

(3) 巩固微型蒸馏操作、微量液体的洗涤、干燥、过滤等操作,学习减压蒸馏操作。

【原理】

反应式：

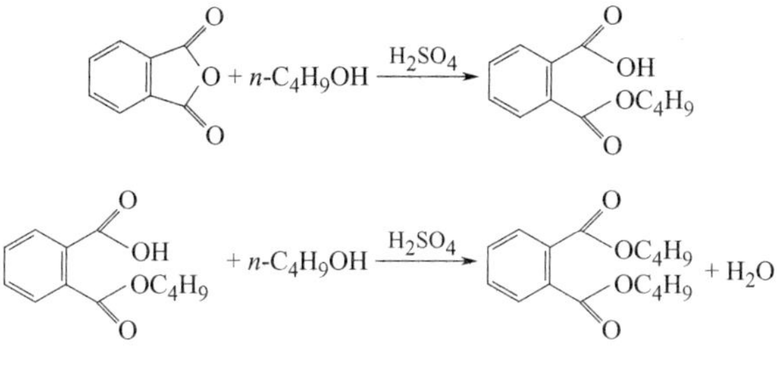

$$\text{邻苯二甲酸酐} + n\text{-}C_4H_9OH \xrightarrow{H_2SO_4} \text{邻苯二甲酸单丁酯}$$

$$\text{邻苯二甲酸单丁酯} + n\text{-}C_4H_9OH \xrightarrow{H_2SO_4} \text{邻苯二甲酸二丁酯} + H_2O$$

【仪器与药品】

(1) 仪器：10mL 二口烧瓶，10mL 圆底烧瓶，球形冷凝管，分水器，直形冷凝管，分液漏斗，H 形分馏头，玻璃漏斗。

(2) 药品：邻苯二甲酸酐，正丁醇，浓硫酸，5％碳酸钠溶液，无水硫酸钠，饱和氯化钠溶液。

【步骤】

在 10mL 二口烧瓶中，放入 2.1g 邻苯二甲酸酐、4.6mL 正丁醇和浓硫酸 3 滴，摇匀，按图 1-5 装置仪器。在分水器中加入正丁醇至支口下 1cm 处。小火加热，邻苯二甲酸酐溶解，随后有水-正丁醇蒸出，分水器中逐步有水分出，回流至分水器中水不再增加（温度升到 140℃[1]）时，停止反应。冷却至 70℃以下，用 7.5mL 5％碳酸钠溶液中和，分液，再用饱和氯化钠溶液洗涤 2 次，无水硫酸钠干燥，按图 2-3(a) 进行减压蒸馏，先用水泵进行减压，蒸去正丁醇，再用油泵进行减压蒸馏，收集 180～190℃/1333Pa 的馏分[2]，产量约为 3g。

【附注】

[1] 邻苯二甲酸二丁酯在酸性条件下，当温度超过180℃时易发生分解反应：

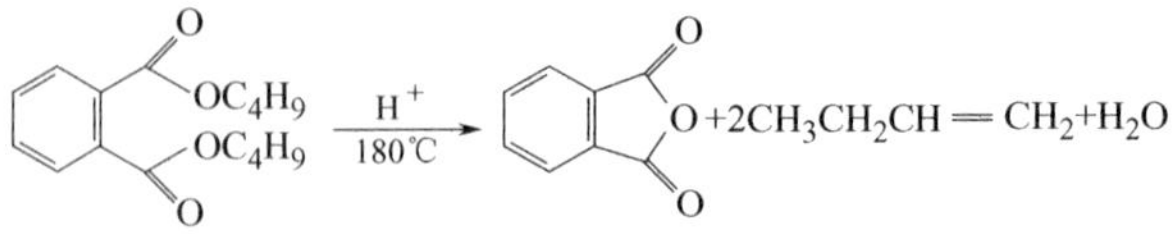

$$\text{邻苯二甲酸二丁酯} \xrightarrow[180℃]{H^+} \text{酐} + 2CH_3CH_2CH=CH_2 + H_2O$$

[2] 根据真空度的不同，可收集 165～170℃/0.27kPa，175～180℃/0.67kPa，200～210℃/2.7kPa 的馏分。

实验 24　丁二酸酐

【目的与要求】

掌握酸酐的制备方法。

【原理】

反应式：

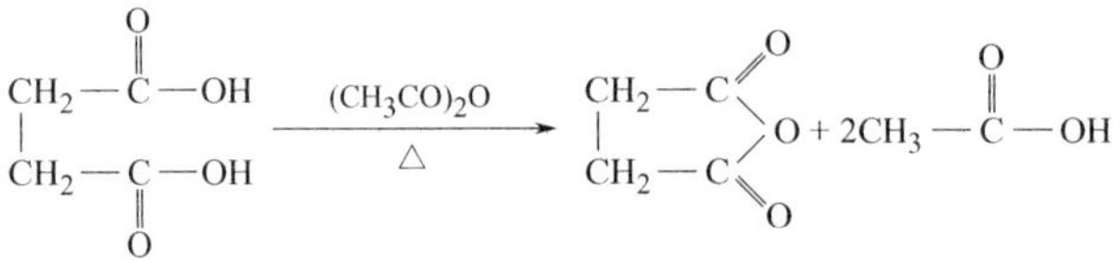

$$\begin{array}{c} CH_2-COOH \\ | \\ CH_2-COOH \end{array} \xrightarrow[\Delta]{(CH_3CO)_2O} \begin{array}{c} CH_2-C \\ | \\ CH_2-C \end{array}\!\!\!\!\!\!>\!\!O + 2CH_3-C-OH$$

反应机理表示如下：

（反应机理示意图）

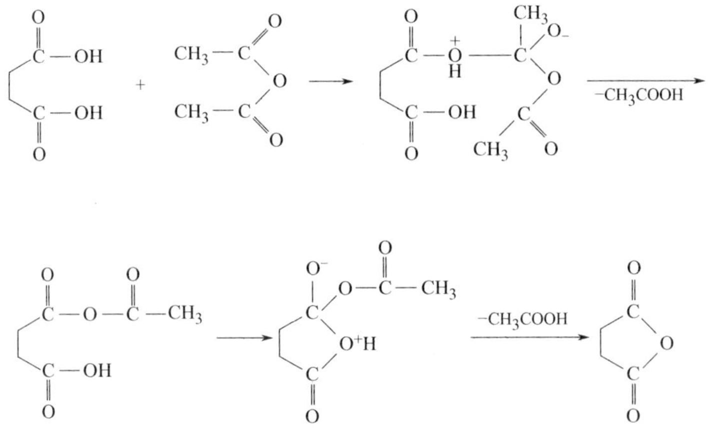

【仪器与药品】

（1）仪器：圆底烧瓶，球形冷凝管，干燥管，漏斗，吸滤瓶。

（2）药品：丁二酸，乙酐，乙醚，无水氯化钙。

【步骤】

在干燥的圆底烧瓶里加入 2g 丁二酸和 3.2mL 新蒸馏过的乙酐，装上球形冷凝管及无水氯化钙干燥管[1]。在沸水浴上加热，间歇摇荡，待丁二酸完全溶解成澄清溶液后，再继续加热 1h，以促使反应完全。

移去水浴，反应物用冷水浴冷却，可见到晶体析出，再用冰水浴充分冷却，抽滤，用玻璃瓶塞将粗产物中的液体挤压出去，再用乙醚洗涤 2 次，每次用 2mL。晶体用红外灯干燥，产量约为 1.1g。得白色针状结晶，测定熔点。

纯丁二酸酐为无色针状晶体，熔点为 119.6℃。其红外光谱如图 3-9 所示。

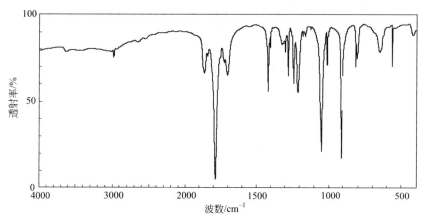

图 3-9　丁二酸酐的红外光谱图

【附注】

[1] 本实验所用仪器必须事先干燥。

思　考　题

还可以用什么方法从丁二酸制丁二酸酐？

实验 25　苯甲酸乙酯

【目的与要求】

(1) 掌握酯化反应原理及苯甲酸乙酯的制备方法，了解三元共沸除水原理。

(2) 复习分水器的使用及液体有机化合物的精制方法。

(3) 学习减压蒸馏提纯有机化合物的方法，进一步掌握蒸馏、萃取、干燥等基本操作。

【原理】

苯甲酸、乙醇在浓硫酸的催化下进行酯化反应，生成苯甲酸乙酯与水。

$$\text{C}_6\text{H}_5\text{COOH} + \text{C}_2\text{H}_5\text{OH} \xrightarrow{\text{H}_2\text{SO}_4} \text{C}_6\text{H}_5\text{COOC}_2\text{H}_5 + \text{H}_2\text{O}$$

【仪器与药品】

(1) 仪器：圆底烧瓶，球形冷凝管，分水器，分液漏斗，锥形瓶，直形冷凝管，烧杯，温度计。

(2) 试剂：苯甲酸，无水乙醇，浓硫酸，碳酸钠，环己烷，石油醚，无水硫酸镁，沸石。

【步骤】

50mL 圆底烧瓶中加入 2g 苯甲酸、5mL 乙醇、4mL 环己烷、1.5mL 浓硫酸，沸石，充分摇匀。按照实验仪器图 1-3 装置仪器，加热至液体回流。开始时回流要慢。随着回流的进行，分水器中出现上下两层。当下层接近分水器支管时将下层液体放出至烧杯中。回流至分水器中水不再增加，停止反应。按图 2-1(a) 装置仪器（或使用旋转蒸发仪），加热蒸出过量的乙醇和环己烷,至瓶内有白烟或回流下来液体不呈现出滴状,停止加热,冷却。将反应液转移至分液漏斗中,用饱和碳酸钠溶液洗涤至呈中性(或弱碱性,pH试纸检验)。分去水层,有机层经无水硫酸镁干燥、过滤、使用蒸馏装置(或旋转蒸发仪)蒸去低沸点馏分后,按图2-3(a)使用油泵进行减压蒸馏,收集苯甲酸乙酯(常压下苯甲酸乙酯沸点210~213℃)。

苯甲酸乙酯沸点与压力的关系如下:

压力	mmHg	10	20	40	60	100
	kPa	1.33	2.60	5.32	7.89	13.3
沸点	℃	86.0	101.4	118.2	129.0	143.0

思 考 题

(1) 本实验采用何种措施提高酯的产率?
(2) 为什么采用分水器除水?
(3) 何种原料过量?为什么?为什么要加苯?
(4) 浓硫酸的作用是什么?常用酯化反应的催化剂有哪些?
(5) 为什么用水浴加热回流?
(6) 在萃取和分液时,两相之间有时出现絮状物或乳浊液,难以分层,如何解决?

实验 26 巯基乙酸铵

【目的与要求】

(1) 掌握巯基乙酸铵的制备方法。
(2) 了解冷烫液和固定液的作用原理、制备原理和方法。

【原理】

人的头发主要由角元纤维所组成,是由二硫化键、盐类键、缩氨酸键、范德华键等组合在一起的聚肽链类物质。其中的二硫化键比较脆弱,化学卷发剂的作用机理是利用化学卷发剂使头发结构中的二硫化键断裂,重新组合出半胱氨酸,而使头发卷曲(如下所示)。最后再用固定液使半胱氨酸重新氧化成胱氨酸,从而使头发达到定型的目的。

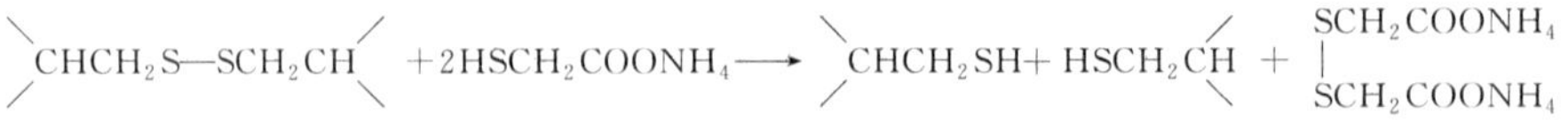

$$\rangle CHCH_2S-SCH_2CH\langle + 2HSCH_2COONH_4 \longrightarrow \rangle CHCH_2SH + HSCH_2CH\langle + \begin{array}{c} SCH_2COONH_4 \\ | \\ SCH_2COONH_4 \end{array}$$

目前我国应用最广泛的冷烫液是含有巯基乙酸铵的复配溶液[1],[2]。其中主要成分巯基乙酸铵的制备方法有多种,下面是较普遍使用的、较容易实现的方法。

$$2ClCH_2COOH + Na_2CO_3 \Longrightarrow 2ClCH_2COONa + H_2O + CO_2$$

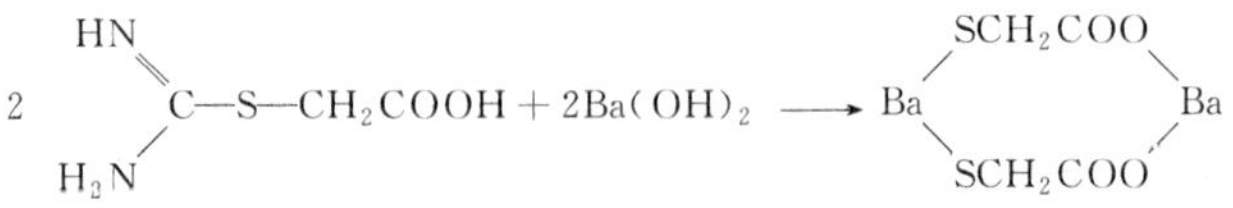

$$ClCH_2COONa + (NH_2)_2CS \Longrightarrow \begin{array}{c} HN \\ \parallel \\ H_2N \end{array} C-S-CH_2COOH + NaCl$$

$$2 \begin{array}{c} HN \\ \parallel \\ H_2N \end{array} C-S-CH_2COOH + 2Ba(OH)_2 \longrightarrow Ba \begin{array}{c} SCH_2COO \\ \\ SCH_2COO \end{array} Ba$$

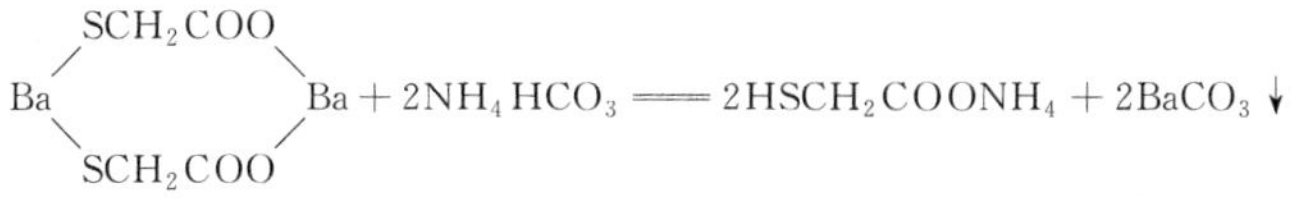

$$\text{Ba}\begin{array}{c}\text{SCH}_2\text{COO}\\\text{SCH}_2\text{COO}\end{array}\text{Ba} + 2\text{NH}_4\text{HCO}_3 =\!=\!= 2\text{HSCH}_2\text{COONH}_4 + 2\text{BaCO}_3\downarrow$$

【仪器与药品】

（1）仪器：三口烧瓶，冷凝管，电磁加热搅拌器，滴液漏斗。

（2）药品：氯乙酸，硫脲，氢氧化钡，碳酸钠，碳酸氢铵。

【步骤】

取 5g 氯乙酸溶解于 3mL 水中，在不断搅拌下向其中慢慢滴加 33％碳酸钠溶液，直到溶液的 pH 值为 7 时。

称取 3.8g 硫脲溶解于 15mL 70℃热水中，在搅拌下加入上面制得的氯乙酸钠溶液，加热至 80℃，搅拌反应 0.5h，过滤生成的白色沉淀，水洗 2 次。

取 15g Ba(OH)$_2$·8H$_2$O 溶解于 38mL 80℃的水中，溶解完毕，加入上述白色沉淀，在 80℃下搅拌 4h。反应完毕后滤出白色钡盐沉淀，用 50℃热水洗涤 2 次，将所得沉淀放入 31mL 23％的碳酸氢铵溶液中搅拌 30min，过滤后可得含量约为 13.2％的巯基乙酸铵水溶液 25mL[3]。

【附注】

[1] 冷烫液的配方见表 3-3。

[2] 固定液的配方见表 3-4。

表 3-3　冷烫液的配方

化　合　物	m/g	化　合　物	m/g
巯基乙酸铵（10％）	7.5	十二烷基苯磺酸钠	0.004
氨水①		水	1.7
三乙醇胺	0.3	香精	适量
羧甲基纤维素	0.002		

① 氨水是调节 pH 值的，将巯基乙酸铵的 pH 值调节至 9.3。

表 3-4　固定液的配方

化　合　物	m/g	化　合　物	m/g
柠檬酸	2	双氧水（30％）	8
酒石酸	1	水	89

[3] 巯基乙酸铵含量可用碘量法测定：

$$2\text{HSCH}_2\text{CO}_2\text{NH}_4 + \text{I}_2 \longrightarrow \begin{array}{c}\text{SCH}_2\text{CO}_2\text{NH}_4\\|\\\text{SCH}_2\text{CO}_2\text{NH}_4\end{array} + 2\text{HI}$$

思　考　题

本实验得到的产品是巯基乙酸铵水溶液，如果要得到固体产品，需要怎样进一步处理？

Experiment 27　Synthesis of Ethyl Acetate

【Experimental Objectives】

(1) Understand the principle of esterification and ester preparation methods.

(2) Learn and master the operations of micro-distillation, reflux and liquid-liquid separation.

【Experimental Principles】

Ethyl acetate (systematically ethyl ethanoate, commonly abbreviated EtOAc or EA) is the organic compound with the formula $CH_3-COO-CH_2-CH_3$. This colorless liquid has a characteristic sweet smell (similar to pear drops) and is used in glues, nail polish removers, decaffeinating tea and coffee, and cigarettes (see list of additives in cigarettes).

Esters may be prepared by direct esterification of an acid, with an alcohol in the presence of an acid catalyst (sulfuric acid, hydrogen chloride) and by alcoholysis of acid chlorides, acid anhydrides, and nitriles. Occasionally they are prepared by heating the metallic salt of a carboxylic acid with an alkyl halide or alkyl sulfate.

Strong acids, typically sulfuric acid, catalyze this reaction. Many other acids are also used. Esterification is highly reversible. The simple reaction of one equivalent each of acid and alcohol gives a mixture of starting materials and products. The yield of the product may be improved using le Chatelier's principle:

(1) using the alcohol in large excess (i.e. as a solvent).

(2) using a dehydrating agent. Sulfuric acid (H_2SO_4) not only catalyzes the reaction, but sequesters water (a reaction product).

(3) removal of water by physical means such as an azeotropic distilation with cyclohexane or toluene.

$$CH_3COOH + CH_3CH_2OH \xrightleftharpoons[\triangle]{H_2SO_4} CH_3COOC_2H_5 + H_2O$$

$$2CH_3CH_2OH \xrightleftharpoons[\triangle]{H_2SO_4} C_2H_5OC_2H_5 + H_2O$$

【Apparatus and Reagents】

(1) Apparatus: Heating mantle, round bottom flask, straight and spherical condensers, distillation head, H type fractionator, separatory funnel, conical flask, thermometer.

(2) Reagents: Glacial acetic acid, anhydrous ethanol, concentrated sulfuric acid, zeolite, saturated aqueous sodium carbonate, anhydrous magnesium sulfate, blue litmus paper.

【Experimental Procedure】

In a round-bottomed flask provided with a reflux condenser, mix thoroughly 3.8mL (0.066mol) of glacial acetic acid, 6mL (0.1mol) of 95% ethanol, and 0.8mL of concentrated sulfuric acid (Caution: corrosive!). Add a small amount of zeolite, heat the mixture to boiling, and continue heating under reflux for 20min. Cool the contents of the flask slightly, remove the condenser, and set it downward for distillation.

Distill the reaction mixture until the residue in the distilling flask amounts to only a few milliliters[1]. Place the distillate[2] in a 25mL conical flask or beaker, add carefully in small portions of saturated aqueous Na_2CO_3 (Caution: foaming!), until no bubbles are produced and testing with blue litmus paper shows that the acid present is neutralized completely[3].

Transfer the mixture to a separatory funnel, remove the lower layer as completely as possible, and wash the ethyl acetate layer with 5mL of saturated aqueous NaCl, 5mL of saturated aqueous $CaCl_2$ and 5mL of water, respectively. Separate the water carefully and dry the ethyl acetate with a small amount of anhydrous $MgSO_4$[4].

Filter through a funnel into a dry distilling flask. Add a small amount of zeolite, distill and collect the boiling portion at 74—77℃ in a weighed bottle[5]. The yield is 1.0—1.3g.

【Notes】

[1] Avoid overheating the residue.

[2] The distillate consists of ethyl acetate, together with ethanol, acetic acid, sulfurous acid, and water.

[3] The acid is easily removed by base extraction.

[4] Anhydrous calcium chloride is not suitable because it forms addition complexes with esters.

[5] Ethyl acetate can form binary azeotrope and trinary azeotrope with water and ethanol. Therefore, in the presence of water and ethanol, the boiling point of the product will decrease, which affects the yield.

【Thinking Questions】

(1) What is the purpose of adding sulfuric acid in the preparation of ethyl acetate? Would any of the ester be formed in the absence of sulfuric acid?

(2) What procedures may be used to drive an esterification toward completion?

(3) Why the reaction mixture would be washed by saturated aqueous Na_2CO_3 solution, saturated aqueous NaCl solution, saturated aqueous $CaCl_2$ solution? Can we wash the reaction mixture with saturated aqueous $CaCl_2$ solution first? Why?

Experiment 28　Synthesis of Acetanilide

【Experimental Objectives】

(1) Learn the principle and operation of aniline acetylation.

(2) Master the operations of fraction and recrystallization.

【Experimental Principles】

Acetylation is often used to "protect" a primary or secondary amine functional group. Acetylated amines are less susceptible to oxidation, less reactive in aromatic substitution reactions, and less prone to participate in many of the typical reactions of free amines, since they are less basic. The amino group can be regenerated readily by hydrolysis in acid or base.

Arylamines (and aliphatic amines) may be acetylated by means of acetic anhydride or acetyl chloride or by heating the amine with glacial acetic acid under conditions that permit removal of the water formed in the reaction. The last procedure is an economical one but requires a relatively long period of heating.

$$PhNH_2 + CH_3COOH \rightleftharpoons PhNHCOCH_3 + H_2O$$

【Apparatus and Reagents】

(1) Apparatus: Heating mantle, round-bottomed flask, beaker, Vigreux fractionating column, distillation head, thermometer, conical flask, Buchner funnel, suction flask.

(2) Reagents: Aniline, glacial acetic acid.

【Experimental Procedure】

In a 50mL round-bottom flask, place 2.0mL (0.022mol) of aniline[1], 3.0mL (0.051mol) of glacial acetic acid. Provide the flask with a Vigreux fractionating column fitted with a thermometer and arranged for distillation. Heat the flask gently, so that the solution boils quietly and the vapor does not rise into the column.

After 20min, increase the heating slightly so that the water formed in the reaction, together with a little acetic acid, distills over very slowly at a uniform rate (vapor temperature 104—105℃). When the water formed in reaction and most of acetic acid has been distilled off, head temperature will drop. Stop heating and pour the reaction mixture out at once into about 30mL of cold water in a large beaker. Stir the aqueous mixture vigorously to avoid formation of large lumps of the product. Collect the acetanilide with suction, wash with a little cold water, and press it firmly on the filter paper.

Crystallize the moist product from hot water (about 20mL/g) with addition of about 0.5g of decolorizing activated carbon. For filtration use a large fluted filter and a large funnel with a short, wide stem. Cool the filtrate rapidly while stirring vigorously to obtain small crystals. Allow the material to stand for about 10min in an ice-water bath and then collect the crystals with suction. Wash the product with a small amount of cold water and spread it on a clean watch glass to dry. The yield is about 0.4g[2] and the melting point is 113—114℃ (Literature: 114.3℃).

【Notes】

[1] The aniline stored for a long time is dark colored. It had better be distilled by vacuum distillation before use because the colored impurity may affect the quality of acetanilide.

[2] If the material is dark colored it should be recrystallized.

【Thinking Questions】

(1) When acetic acid is used for acetylation of an amine, why is it desirable to use an excess of the acid and to distill off the water formed in the reaction?

(2) In the preparation of p-nitroaniline, why is aniline converted to acetanilide before nitration?

(3) Try to summarize the application of acetanilide in organic synthesis.

3.8 芳香族硝基化合物

实验29 硝 基 苯

【目的与要求】

掌握芳香族化合物硝化反应的原理和方法。

【原理】

主反应：

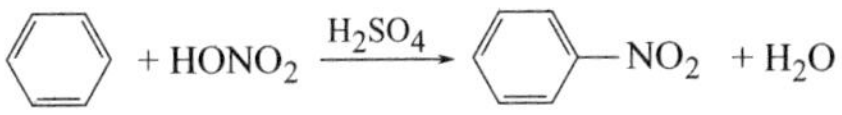

$$\text{C}_6\text{H}_6 + \text{HONO}_2 \xrightarrow{\text{H}_2\text{SO}_4} \text{C}_6\text{H}_5\text{-NO}_2 + \text{H}_2\text{O}$$

副反应：

$$\text{C}_6\text{H}_5\text{-NO}_2 + \text{HONO}_2 \xrightarrow{\text{H}_2\text{SO}_4} \text{C}_6\text{H}_4(\text{NO}_2)_2 + \text{H}_2\text{O}$$

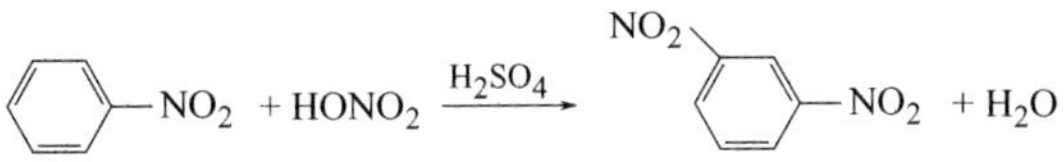

【仪器与药品】

(1) 仪器：二口圆底烧瓶，冷凝管，电磁加热搅拌器，分液漏斗，温度计，回收瓶，锥形瓶。

(2) 药品：苯，硝酸，浓硫酸，10％碳酸钠溶液，饱和食盐水，无水氯化钙。

【步骤】

在二口圆底烧瓶上口装回流冷凝管，侧口插装温度计，其水银球接近瓶底。在该烧瓶中加入 4.5mL 苯。通过冷凝管上口，在搅拌状态下，将已冷却的混酸[1]分多次加入苯中，必须待反应物的温度不再上升而趋于下降时，才继续加入混酸（为什么？）。维持反应物温度保持在 40~50℃ 之间，必要时可用冷水浴冷却烧瓶。加料完毕后，约于 10min 内把反应混合物的温度加热到 60~65℃，保温反应 30min[2]。

反应毕冷却，将混合物倒入分液漏斗中。静置分层，分出酸层（哪一层？怎样判断和检验？），倒入回收瓶内。粗硝基苯[3]先用等体积的冷水洗涤，再用 10％碳酸钠溶液洗涤，直到洗涤液不显酸性。最后用水洗至中性（如何检验？）。分离出粗硝基苯，放在干燥的小锥形瓶中，加入无水氯化钙干燥，间歇振荡锥形瓶。

把澄清透明的硝基苯倒入蒸馏烧瓶中以空气冷凝管蒸馏，收集 204~210℃ 的馏分，产量约为 4g。为了避免残留在烧瓶中的二硝基苯在高温下分解而引起爆炸，注意切勿将产物蒸干。

纯硝基苯为无色液体，具有苦杏仁气味，沸点为 210.9℃。

【附注】

[1] 混酸配制法：将锥形瓶放入冷水浴中加入 5mL 浓硫酸，边摇动锥形瓶边将 3.7mL 硝酸慢慢地注入浓硫酸中，混合均匀即成。

[2] 用吸管吸取少许上层反应液，滴到饱和食盐水中，当观察到油珠下沉时，表示硝化反应已经完成。

[3] 硝基苯有毒，处理时须加小心。如果溅在皮肤上，可先用少量酒精洗擦，再用肥皂水洗净。

思 考 题

(1) 硫酸在本实验中起什么作用？

(2) 如果一次就将混酸加完会发生什么结果？

(3) 硝化反应温度过高有什么影响？

(4) 粗品蒸馏时为什么不能蒸干？

实验 30　间二硝基苯

【目的与要求】

(1) 进一步掌握硝化反应和多硝基化合物的制备方法。

(2) 进一步理解定位效应。

【原理】

主反应：

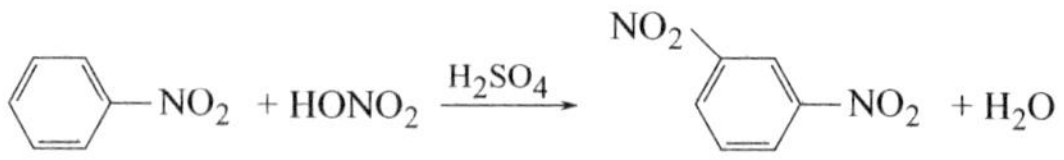

$$\text{C}_6\text{H}_5\text{-NO}_2 + \text{HONO}_2 \xrightarrow{\text{H}_2\text{SO}_4} \text{m-C}_6\text{H}_4(\text{NO}_2)_2 + \text{H}_2\text{O}$$

副反应：

$$\text{C}_6\text{H}_5\text{-NO}_2 + \text{HONO}_2 \xrightarrow{\text{H}_2\text{SO}_4} \text{p-O}_2\text{N-C}_6\text{H}_4\text{-NO}_2 + \text{o-C}_6\text{H}_4(\text{NO}_2)_2$$

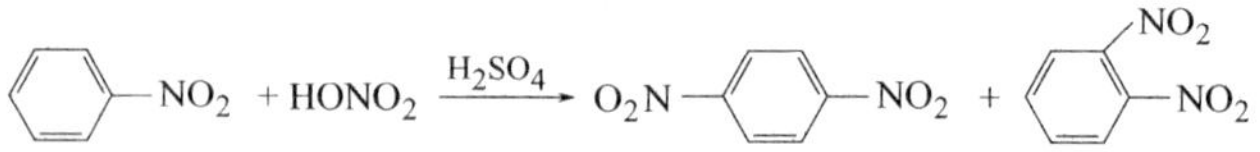

【仪器与药品】

(1) 仪器：圆底烧瓶，空气冷凝管，磁力加热搅拌器，漏斗，吸滤瓶，烧杯。
(2) 药品：硝基苯，硝酸，浓硫酸，碳酸钠，95%乙醇。

【步骤】

在干燥的圆底烧瓶中加入4.5mL浓硫酸。把烧瓶置于冷水浴中，慢慢地加入3.2mL硝酸，同时开动磁力加热搅拌器搅拌，然后加入2mL硝基苯。在烧瓶上装空气冷凝管。加热搅拌反应1h，至反应完全[1]。

当反应混合物冷却到约70℃时，在剧烈搅拌下，把反应物以细流慢慢地倒入盛有40mL冷水的烧杯中[2]。粗二硝基苯成块状物沉入烧杯底。冷却后，倾去酸液。烧杯中加入25mL热水加热至固体熔化。搅拌数分钟后，冷却之，倾去稀酸液。烧杯中再加入25mL热水，加热至固体熔化，然后一边搅拌，一边分几次加入粉状碳酸钠，直到水溶液呈显著碱性为止（用什么试纸检验？）。冷却后，倾去碱液。粗二硝基苯再用50mL热水分2次洗涤。冷却后减压过滤，尽量挤压去水分。取出产物[3]。用95%乙醇进行重结晶，产量约为2g。

纯间二硝基苯为无色针状晶体，熔点为90.02℃。

【附注】

[1] 用吸管吸取少许上层反应物，滴入盛冷水的试管中，如果立刻有淡黄色的固体析出，表示反应已经完成，为什么？如果呈半固体状，则需继续加热，直到反应完成为止。
实验在三口烧瓶中利用机械搅拌进行操作。硝化操作在通风橱内进行。
[2] 铁夹夹紧，摆动时注意勿与连接处松开。此步操作应在通风橱内进行。
[3] 硝基苯有毒，操作时必须小心，勿使其接触皮肤。

思 考 题

(1) 为什么制备间二硝基苯要在较强烈的反应条件下进行？
(2) 进行硝化反应时最后通常是将反应后混合物倒入大量水中，这步操作的目的何在？
(3) 制得的间二硝基苯有什么杂质？如何除掉？

Experiment 31　Synthesis of 2-Nitroresorcinol

【Experimental Objectives】

(1) Study the localization of the electrophilic substitution reaction on benzene ring and the application of the sulfonation reaction.

(2) Learn the synthesis of 2-nitro-1,3-benzenediol.

【Experimental Principles】

In the course of certain organic synthesis it is often advantageous to introduce a group that will effectively block or protect certain potentially reactive sites on a molecule from attack by some specific reagent. Important features of such a group are that it should be easily introduced and easily removed after some crucial step in the synthesis has been performed.

In the process of converting resorcinol (1,3-dihydroxybenzene) to 2-nitroresorcinol, the starting compound is first sulfonated to give resorcinol-4,6-disulfonic acid in which two of the three positions most susceptible to nitration are now blocked. Nitration followed by steam distillation of an acidic solution of the nitrated disulfonic acid to remove the sulfonic acid groups results in pure 2-nitroresorcinol.

【Apparatus and Reagents】

(1) Apparatus: Heating mantle, Beaker, glass rod, dropping funnel, round bottom flask, H-type fractional head, straight condenser, conical flask, Buchner funnel.

(2) Reagents. Powdered resorcinol, concentrated sulfuric acid (98%, $d=1.84$), mixed acid (nitric acid 70%—72%, $d=1.42$), concentrated sulfuric acid (98%, $d=1.84$), 95% ethanol, ice.

【Experimental Procedure】

Place 1.54g (0.014mol) of powered resorcinol[1] in a 50mL beaker and add 5.6mL (10.08g, 0.103mol) of concentrated sulfuric acid (98%, $d=1.84$). If a thick slurry of the 4,6-disulfonic acid does not form in a few minutes, warm the mixture to 60—65℃. Allow the slurry to stand for 15min.

Prepare a mixture of 0.88mL (0.876g, 0.0139mol) of nitric acid (70%—72%, $d=1.42$) and 1.24mL (2.38g, 0.0232mol) of concentrated sulfuric acid, and cool it in an ice bath. Cool the slurry in an ice-salt bath to a temperature of 5—10℃, stir it, and slowly (dropwise) add the cold acid solution from a addition funnel suspended over the beaker. The temperature of the reaction mixture should not exceed 20℃. After the yellowish mixture has stood for 15min, it should be cautiously diluted with 10g of crushed ice so that the temperature never exceeds 50℃.

Transfer the mixture to a 50mL round-bottomed flask, add 0.02g of urea[2], and carry out an indirect steam distillation until no more of the orange-red, solid 2-nitroresorcinol (m.p. 215℃) appears. The product will usually appear after about 5min of steam distillation. The product may not steam-distill if too much steam has condensed in the distillation flask. In this event shut off the steam and heat the flask (a Bunsen burner or a heating mantle will be required) until sufficient water is removed to increase the sulfuric acid concentration in the flask to the point at which desulfonation will occur and product will again distill. If the condenser becomes filled with solidified product, turn off the cooling water for a few minutes until the product has melted and flowed into the receiver.

Cool the distillate in an ice bath and filter it with suction. Recrystallize the product from dilute aqueous ethanol by first dissolving it in 95% ethanol (*ca*. 3mL per gram of product), filtering in hot water slowly until cloudiness (or small amounts of precipitate) persists. And allowing the solution to cool slowly. The yield of 2-nitroresorcinol, is 0.5—0.7g. The melting point of pure 2-nitroresorcinol is 84—85℃.

【Notes】

[1] In order to be sulphonated completely, the resorcinol should have been previously ground to fine power in a mortar.

[2] When urea is added, the salt of the urea and the nitric acid is formed. As the salt dissolves in water, the excess of the nitric acid is removed.

【Thinking Questions】

Why might sulfonation be expected to occur most readily in the 4-and 6-positions father than the 2-position? How does the overall reactivity (toward electrophilic attack) of the disulfonic acid compare with that of resorcinol?

3.9 胺及季铵盐

实验32 间硝基苯胺

【目的与要求】

掌握多硝基苯的部分还原反应。

【原理】

反应式：

$$Na_2S + S \rightleftharpoons Na_2S_2$$

$$O_2N\text{-}C_6H_4\text{-}NO_2 + Na_2S_2 + H_2O \longrightarrow O_2N\text{-}C_6H_4\text{-}NH_2 + Na_2S_2O_3$$

【仪器与药品】

(1) 仪器：二口圆底烧瓶，电磁加热搅拌器，冷凝管，漏斗，吸滤瓶，锥形瓶。

(2) 药品：硫化钠，硫黄粉，间二硝基苯，盐酸，氨水。

【步骤】

(1) 多硫化钠溶液制备[1]：在锥形瓶中加入2g硫化钠晶体，8mL水和0.5g硫黄粉。边振荡搅拌边加热，直到硫黄粉全部溶解。过滤除去不溶物得澄清溶液，冷却备用。

(2) 间硝基苯胺的制备：在装有滴液漏斗和回流冷凝管的二口圆底烧瓶中加入1.25g间二硝基苯和10mL水。在电磁加热搅拌器的搅拌下加热到回流，使熔融的间二硝基苯与水充分混合成很细的悬浮液。用滴管或滴液漏斗把多硫化钠溶液在加热和搅拌下滴入烧瓶中，继续加热回流反应20min[2]。

静置并迅速冷却反应混合物。将析出的粗间硝基苯胺减压过滤,除去水分。用少量冷水洗涤滤饼以除去残留的硫代硫酸钠。取出粗产物,加入稀盐酸(由8mL水和2mL浓盐酸配制),加热使间硝基苯胺成盐溶解。抽滤除去硫黄和未反应的间二硝基苯。在滤液中加入过量浓氨水(使pH=8),至溶液中析出黄色的间硝基苯胺。减压过滤除去碱液,再用冷水洗涤到中性,抽滤,干燥后得粗产品。粗产品用水重结晶,得黄色晶体。

纯间硝基苯胺为浅黄色针状晶体,熔点为114℃。

【附注】

[1] 也可用 Na_2S,$(NH_4)_2S$,NH_4HS 作部分还原剂。

[2] 反应终点的判断:用胶头滴管吸少量反应物,滴在用硫酸铜溶液浸过的滤纸上,若生成的硫化铜黑色斑点于20s内不消失,即可认为反应已达终点。

思 考 题

(1) 能否用铁和盐酸做还原剂来制取间硝基苯胺?

(2) 产品中混有哪些杂质?怎样除去?

实验33 邻氨基苯甲酸

【目的与要求】

(1) 了解 Hofmann 酰胺降解制备伯胺的原理。

(2) 巩固微型回流、过滤等基本操作。

【原理】

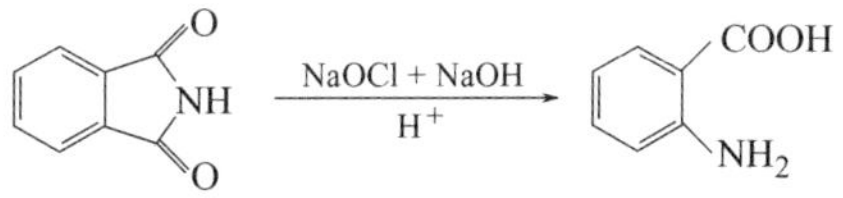

【仪器与药品】

(1) 仪器:10mL 二口圆底烧瓶,球形冷凝管,干燥管,玻璃漏斗,电磁加热搅拌器。

(2) 药品:邻苯二甲酰亚胺,次氯酸钠,氢氧化钠,浓盐酸,冰醋酸。

【步骤】

9mL 5.25%的次氯酸钠溶液、3mL 25%氢氧化钠溶液加入装有电磁加热搅拌器的二口圆底烧瓶中,冰水冷却至5℃以下,加入邻苯二甲酰亚胺0.9g,剧烈搅拌,再加入2mL 25%氢氧化钠溶液,继续搅拌。当温度不再上升时加热到80℃,并在此温度下保持几分钟,冷却到室温,并在室温放置10min。冰水冷却,在搅拌下加入3mL浓盐酸,使其pH值控制在10。缓慢加入1mL冰醋酸,充分冷却,使沉淀完全析出,抽滤,冰水洗涤至晶体无醋酸气味,干燥测定熔点。邻氨基苯甲酸为白色晶体,熔点145℃,能升华。

思 考 题

请写出本实验的反应机理。

实验34 溴化四丁基铵

【目的与要求】

(1) 学习季铵盐的制备原理和合成方法。

(2) 巩固微型回流、过滤等基本操作。

【原理】

反应式：

$$(C_4H_9)_3N + C_4H_9Br \longrightarrow (C_4H_9)_4NBr$$

【仪器与药品】

(1) 仪器：10mL圆底烧瓶，球形冷凝管，干燥管，玻璃漏斗。

(2) 药品：三丁胺，溴丁烷。

【步骤】

在10mL圆底烧瓶中，放入4.1mL三丁胺、2.5mL溴丁烷、沸石1粒，按图1-2(b)装置仪器[1]。回流反应1h。停止加热，冷却。待固体产物析出后抽滤，尽量抽去液体，干燥，称重，产量约为1.4g。

【附注】

[1] 本实验所用仪器必须干燥，同时在冷凝管上要安装干燥管。

思 考 题

为什么本实验所用的仪器必须干燥？

实验35　溴化四乙基铵

【目的与要求】

(1) 学习季铵盐的制备原理和合成方法。

(2) 巩固微型回流、过滤等基本操作。

【原理】

反应式：

$$(C_2H_5)_3N + C_2H_5Br \longrightarrow (C_2H_5)_4NBr$$

【仪器与药品】

(1) 仪器：10mL圆底烧瓶，球形冷凝管，干燥管，玻璃漏斗。

(2) 药品：三乙胺，溴乙烷。

【步骤】

在10mL圆底烧瓶中，放入4.1mL三乙胺、2.5mL溴乙烷、沸石1粒，按图1-2(b)装置仪器[1]。回流反应1h[2]。

停止加热，冷却。待固体产物析出后抽滤，尽量抽去液体，干燥，称重，产量约为2.4g。

【附注】

[1] 本实验所用仪器必须干燥，同时在冷凝管上要安装干燥管。

[2] 也可以将两种反应物混合后，密闭放置1周。

Experiment 36　Synthesis of Benzyltriethylammonium Chloride

【Experimental Objectives】

(1) Learn the preparation principle and method of quaternary ammonium salts.

(2) Master the operations of micro-scale reflux and filtration.

【Experimental Principles】

Benzyltriethylammonium chloride is a phase-transfer catalyst or PTC. A catalyst that facilitates the migration of a reactant from one phase into another phase where reaction occurs. Phase-transfer catalysis is a special form of heterogeneous catalysis. Ionic reactants are often soluble in an aqueous phase but insoluble in an organic phase in the absence of the phase-transfer catalyst. The catalyst functions like a detergent for solubilizing the salts into the organic phase. Phase-transfer catalysis refers to the acceleration of the reaction upon the addition of the phase-transfer catalyst.

Benzyltriethylammonium chloride could be prepared by the reaction of benzyl chloride with triethylamine.

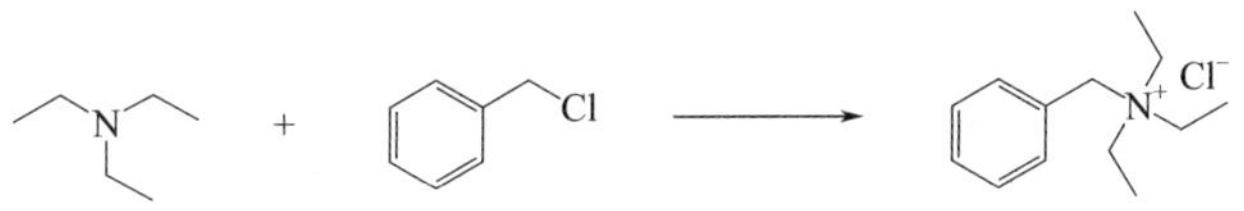

【Apparatus and Reagents】

(1) Apparatus: Heating mantle, round bottom flask, allihn condenser, drying tube, glass funnel.

(2) Reagents: Triethylamine, benzyl chloride, zeolite.

【Experimental Procedure】

In a 25mL round bottomed flask, add triethylamine (2.7mL), benzyl chloride (2.8mL) and a small amount of zeolite. Set up a dry and reflux reaction apparatus according to Figure 1-2(b). Heat the reactants mixture to reflux for 1h[1], then stop heating, cool down and filtrate after precipitation. Try to remove more liquid during filtration, then dry and weigh the solid product. The yield is about 0.8g.

【Notes】

[1] The reaction temperature should not be controlled.

3.10 重氮盐及其反应

实验37 氯 苯

【目的与要求】

(1) 掌握重氮化反应的原理和重氮盐的制备。

(2) 掌握 Sandmeyer 反应制备卤代苯的方法。

【原理】

主反应：

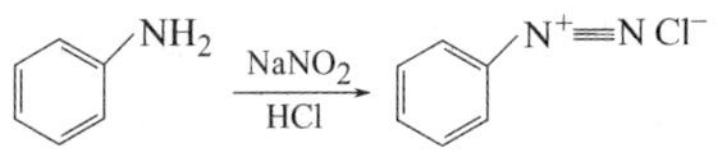

$$2CuSO_4 + 2NaCl + NaHSO_3 + 2NaOH \longrightarrow 2CuCl + Na_2SO_4 + NaHSO_4 + H_2O$$

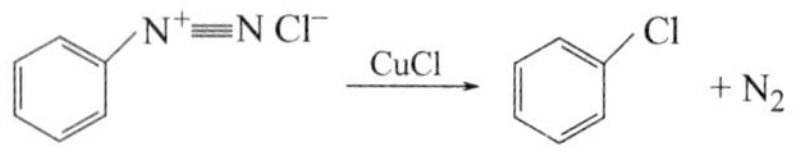

$$\text{C}_6\text{H}_5\text{N}_2^+\text{Cl}^- \xrightarrow{\text{CuCl}} \text{C}_6\text{H}_5\text{Cl} + \text{N}_2$$

副反应：

$$\text{C}_6\text{H}_5\text{N}_2^+\text{Cl}^- + \text{H}_2\text{O} \xrightarrow{\text{H}^+} \text{C}_6\text{H}_5\text{OH} + \text{HCl} + \text{N}_2$$

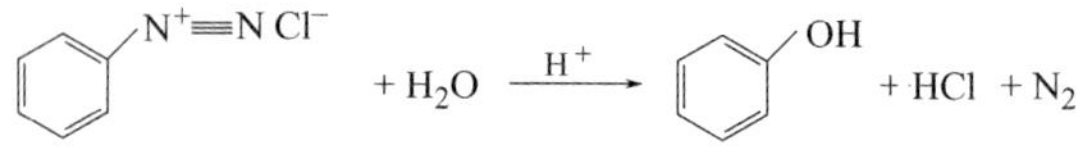

【仪器与药品】

（1）仪器：圆底烧瓶，电磁加热搅拌器，冷凝管，微型蒸馏头，滴管，分液漏斗。

（2）药品：硫酸铜，苯胺，亚硝酸钠，浓盐酸，氯化钠，亚硫酸氢钠，氢氧化钠，浓硫酸，无水氯化钙。

【步骤】

（1）氯化亚铜的制备：将 2.17g 硫酸铜（$\text{CuSO}_4 \cdot 5\text{H}_2\text{O}$）和 0.67g 氯化钠溶于 6.5mL 水中，加热至 60～70℃得溶液（A）。将 0.8g 亚硫酸氢钠和 0.5g 氢氧化钠溶于 3.0mL 水中，加热至 60～70℃得溶液（B）。趁热合并溶液（A）和溶液（B）[1]，溶液即析出白色粉末状固体[2]。冷却至室温后移去上层溶液，剩余固体用水洗涤 2 次，得白色粉末状的氯化亚铜。在搅拌下，将氯化亚铜溶于盛有 3.5mL 浓盐酸的圆底烧瓶中，密闭并置于冰盐浴中冷却。

（2）重氮盐的制备：在圆底烧瓶中依次加入 0.6mL 苯胺、1g 冰、0.5mL 水和 2mL 浓盐酸，搅拌使苯胺完全溶解。将圆底烧瓶置于冰盐浴中迅速降温至 0℃，此时得到苯胺盐酸盐糊状物[3]。在另外的圆底烧瓶中，将 0.53g 亚硝酸钠溶于 1.5mL 水中并冷却至 0℃。在搅拌并保持低温下（温度不超过 5℃），用滴管将亚硝酸钠溶液慢慢加入到苯胺盐酸盐中[4]。当加入 85%～90%的亚硝酸钠溶液时即用碘化钾-淀粉试纸检验亚硝酸钠是否过量[5]。若试纸变蓝，则重氮化反应已完成。重氮化反应过程中还应用试纸检验，使反应液始终保持呈酸性[6]。

（3）氯苯的制备：将重氮盐溶液慢慢倒入已冷却至 0℃的氯化亚铜盐酸溶液中[7]，并时时振荡，反应物温度保持在 15℃以下，此时有深红色悬浮物析出（重氮盐-氯化亚铜复盐）。将圆底烧瓶在冰浴中摇动 2min，然后在室温下摇动使其接近室温后，在水浴中慢慢升温至 50℃，直至无气泡逸出为止。

将反应混合物进行微型水蒸气蒸馏，直至馏出液中无油珠为止。馏出液倒入微型分液漏斗，分出粗氯苯，依次用等体积的浓硫酸[8]和水洗涤 1 次，分离出氯苯，用无水氯化钙干燥后进行蒸馏，收集 130～133℃的馏分，产量约为 0.5g。

纯氯苯是无色液体，沸点为 132℃，相对密度为 1.107。

【附注】

[1] 在此温度下制得的氯化亚铜质量较好，颗粒较粗，易于漂洗处理。

[2] 若溶液呈蓝绿色则表示还原不完全，仍有二价铜存在，应适当多加亚硫酸氢钠溶液使溶液呈现浅绿色或无色。若溶液呈黄褐色则表示铜以氢氧化亚铜形式存在，应适当加几滴稀盐酸使其转化为氯化亚铜。但酸量要控制好，因为氯化亚铜可溶解于酸中。

[3] 迅速冷却可以使析出的晶体颗粒很细，有利于重氮化反应的进行。

[4] 重氮盐易分解，所以重氮化反应一定要在低温下进行。

[5] 过量的亚硝酸可以把碘化钾氧化成碘从而使淀粉变蓝。近终点时，重氮化反应进行较慢。应加几滴亚硝酸钠溶液，搅拌几分钟后再检验终点。过量的亚硝酸用尿素水溶液分解。

[6] 氯化亚铜在空气中易被氧化且重氮盐久置也会分解，所以两者制备后应立即混合反应。

[7] 若加料过快会有较多副产物偶氮苯等生成，所以加料要慢。

[8] 浓硫酸可除去副产物苯酚和偶氮苯。

思 考 题

(1) 重氮化反应为什么要在低温下进行？亚硝酸过量有什么影响？

(2) 重氮盐溶液和氯化亚铜溶液反应时为什么要保持低温？

(3) 为什么用水蒸气蒸馏法分离产物？

实验38 甲 基 红

【目的与要求】

(1) 掌握重氮化反应和重氮盐偶联反应制取甲基红的方法。

(2) 进一步掌握微型洗涤、过滤、重结晶等基本操作。

【原理】

$$\underset{NH_2}{\underset{|}{C_6H_4}}-CO_2H \xrightarrow{HCl} \underset{NH_3Cl}{\underset{|}{C_6H_4}}-CO_2H \xrightarrow{NaNO_2} \underset{N_2^+}{\underset{|}{C_6H_4}}-CO_2^- \xrightarrow{C_6H_5N(CH_3)_2} \underset{N=N-C_6H_4-N(CH_3)_2}{\underset{|}{C_6H_4}}-CO_2H$$

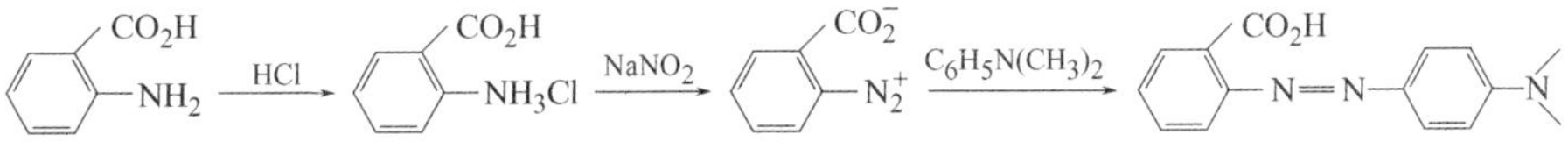

【仪器与药品】

(1) 仪器：30mL 烧杯，电磁加热搅拌器，砂芯漏斗，烧杯。

(2) 药品：邻氨基苯甲酸，浓盐酸，亚硝酸钠，N,N-二甲基苯胺，甲苯，甲醇，乙醇。

【步骤】

在 30mL 烧杯中，加入 1.5mL 邻氨基苯甲酸和 6mL 1∶1 的盐酸，加热使其溶解，放置冷却析出，减压抽滤。将邻氨基苯甲酸盐酸盐加入 20mL 水中溶解，并在冰水浴中冷却到 10℃以下，缓慢将该溶液倒入放在冰水浴里盛有 1.0g 亚硝酸钠的水溶液的烧杯中，得重氮盐溶液。将由 1.5mL N,N-二甲基苯胺、15mL 乙醇组成的溶液加入到上述重氮盐溶液中，并剧烈搅拌。抽滤结晶，并用少量的甲醇洗涤。粗甲基红可用甲苯重结晶，产量约为 0.7g。

思 考 题

试解释甲基红在酸碱介质中的变色原因（用反应式表示）。

实验39 Ⅱ 号 橙

【目的与要求】

(1) 掌握重氮化反应和重氮盐偶联反应制取对位红的方法。

(2) 进一步掌握微型洗涤、过滤、重结晶等基本操作。

【原理】

$$H_2N-C_6H_4-SO_3Na \xrightarrow[HCl]{NaNO_2} N_2^+-C_6H_4-SO_3^-$$

$$N_2^+-C_6H_4-SO_3^- + C_{10}H_6(ONa) \xrightarrow{OH^-} NaO_3S-C_6H_4-N=N-C_{10}H_6(OH)$$

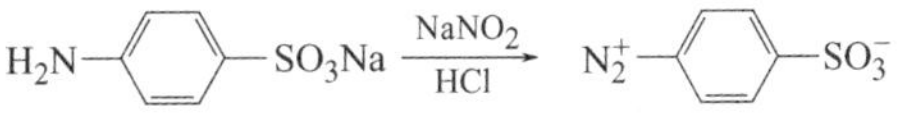

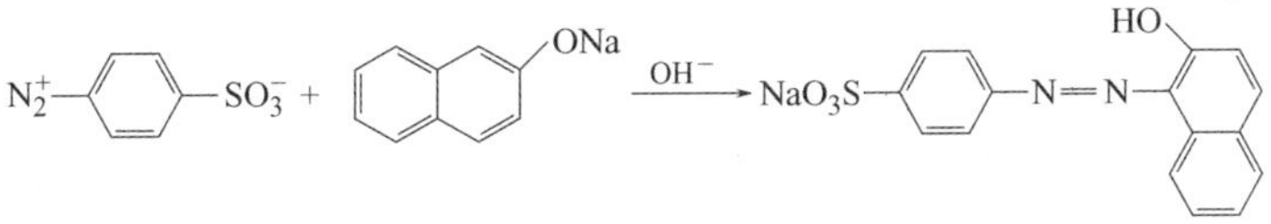

【仪器与药品】

（1）仪器：20mL 锥形瓶，电磁加热搅拌器，砂芯漏斗，30mL 烧杯。

（2）药品：对氨基苯磺酸，浓盐酸，亚硝酸钠，β-萘酚，10%氢氧化钠溶液，饱和食盐水，乙醇。

【步骤】

在锥形瓶中放入 0.24g（1.3mmol）对氨基苯磺酸和 2.5mL 2.5%盐酸溶液，在电磁加热搅拌器的搅拌下溶解，必要时可以加热。冷却后加入 0.1g 亚硝酸钠并搅拌溶解。将此溶液倒入装有 1.3g 冰和 0.3mL 浓盐酸的小烧杯中。片刻有白色粉状重氮盐沉淀析出。在冰浴中冷却备用。

在另一只烧杯中将 0.18g β-萘酚溶于 1mL 冷 10%氢氧化钠溶液中。在搅拌下将制得的重氮盐倒入，有偶联产品析出。加热搅拌溶解，冷却，过滤，用饱和食盐水洗涤。可进一步用乙醇和水的混合溶剂进行重结晶。产量约为 0.3g（带有 2 个结晶水）。

思 考 题

为什么该偶联反应要在氢氧化钠溶液中进行？

Experiment 40 Synthesis of Methyl Orange

【Experimental Objectives】

（1）Learn the method for preparing methyl orange by diazotization reaction and diazo salt coupling reaction.

（2）Master the operations of the washing, filtration and recrystallization.

【Experimental Principles】

Methyl orange is a pH indicator frequently used in titration because of its clear and distinct colour variance at different pH values. Methyl orange shows red colour in acidic medium and yellow colour in basic medium. Because it changes colour at the pH of a mid strength acid, it is usually used in titration for acids. methyl orange does not have a full spectrum of colour change, but it has a sharp end point.

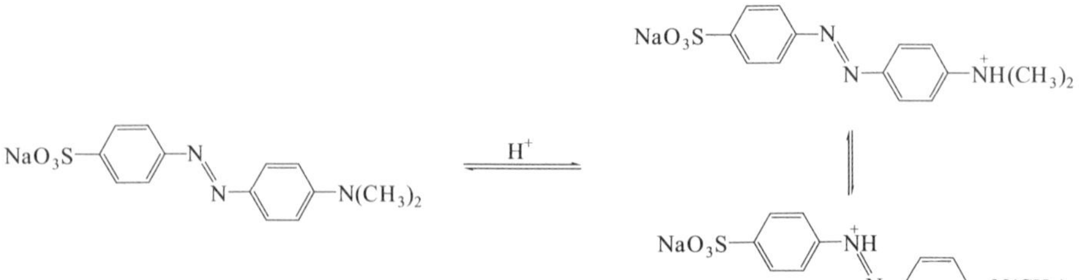

The synthetic method is as follows:

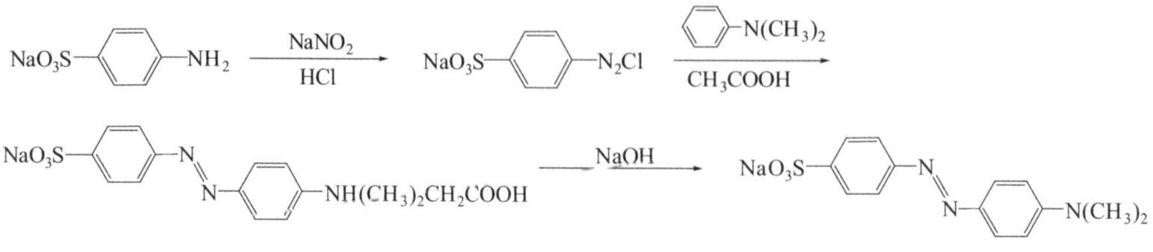

【Apparatus and Reagents】

(1) Apparatus: Round bottom flask, conical flask, electromagnetic heating stirrer, Buchner funnel, beaker.

(2) Reagents: 4-aminobenzenesulfonic acid, concentrated hydrochloric acid, sodium nitrite, N,N-dimethyl aniline, glacial acetic acid, sodium hydroxide, sodium chloride, ethanol, ether.

【Experimental Procedure】

(1) Diazotization reaction: add 4-aminobenzenesulfonic acid (0.3g) and 5% NaOH solution (1.5mL) in a round bottom flask, heat until all solid dissolved in a hot water bath. After cooling down to room temperature, add sodium nitrite (0.12g). After dissolving, the solution mixture is dropwise added into the beaker with 2mL water and 0.4mL HCl. Keep the temperature below 5℃ during the whole process of addition[1]. Then place the mixture in ice bath for 5min.

(2) Coupling reaction: The solution mixed by N,N-dimethyl aniline (0.2mL) and acetic acid (0.15mL) is dropwise added into the diazonium salt solution with occasional stirring. Continue to stir the mixture for another 10min, then the red precipitate of acid yellow is afforded. Then NaOH solution (0.2mL) is added into the mixture and the reaction mixture turns to orange. The precipitate is filtered with a funnel and washed with saturated sodium chloride aqueous solution[2]. The crude product is recrystallized from water and could be washed with a small amount of ethanol and ether[3]. After drying, the final product is weighed and the yield is about 1.2g.

【Notes】

[1] Diazonium salts are unstable, which could be decomposed in high temperature.

[2] Potassium iodide could be oxidized to iodine by the excessive amounts of nitrite salts, which makes the starch turn blue.

[3] The purpose of washing with ethanol and ether is to allow the product to dry quickly.

【Thinking Questions】

Why 4-aminobenzenesulfonic acid should firstly react with sodium hydroxide, then diazotization with sodium nitrite and hydrochloric acid, rather than react directly with sodium nitrite and hydrochloric acid?

3.11 芳香族磺酸

实验41 双酚-S

【目的与要求】

(1) 进一步了解磺化反应。

(2) 巩固微型分馏操作、固体样品的重结晶、干燥等操作。

【原理】

双酚-S 的化学命名为 4,4'-二羟基二苯砜(4,4'-dihydroxyldiphenylsulfone)，与双酚 A 的结构有一定的相似，有着广泛的用途。其合成反应分两步进行。

第一步苯酚的磺化。该反应有邻位和对位产物，在低温时是动力学控制反应，在高温时是热力学控制反应，对位异构体为主产物（如下所示），如果温度再升高，将有间位产物生成。因此反应温度对于提高该反应的产率至关重要。

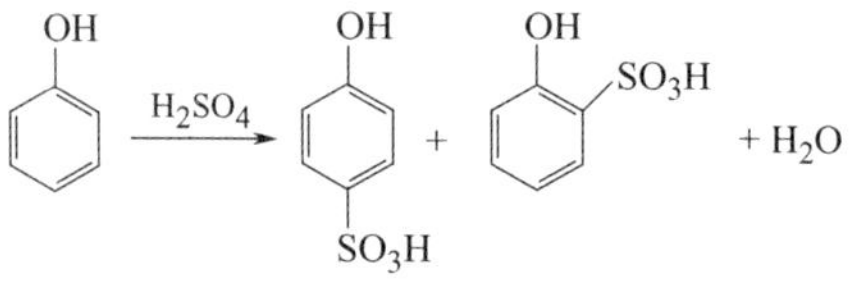

	（I）	（II）
1～5℃	60%	40%
100℃	78.5%	10%

第二步对羟基苯磺酸与另一分子苯酚再磺化，主要副产物是 2,4'-二羟基二苯砜。

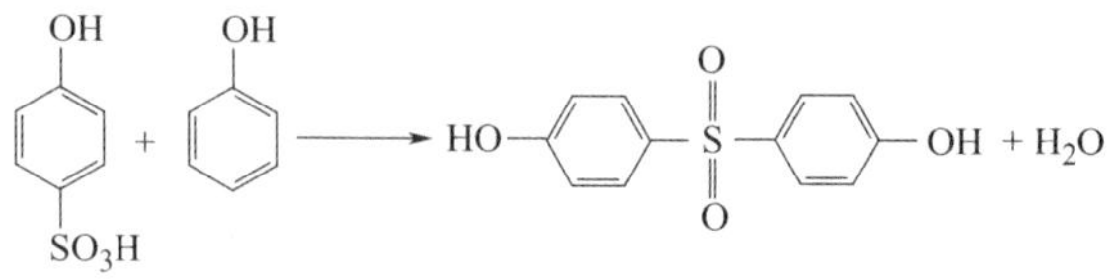

【仪器与药品】

（1）仪器：30mL 三口烧瓶，H 形分馏头，球形冷凝管，砂芯漏斗，温度计，磁力加热搅拌器。

（2）药品：苯酚；浓硫酸；1,2-二氯乙烷；活性炭。

【步骤】

在装有 H 形分馏头、球形冷凝管、温度计和磁力加热搅拌器的 30mL 三口烧瓶中加入 9.4g 苯酚，加热到 110℃左右，在搅拌下慢慢加入 2.7mL 浓硫酸，并使温度维持在 120～150℃，反应 40min。然后降温至 100℃左右，慢慢加入 4mL 1,2-二氯乙烷[1]，继续加热搅拌，此时不断有 1,2-二氯乙烷与水的共沸物蒸出（控制蒸馏速度，不要太快），随着水的不断蒸出，温度不断升高，保持反应温度在 160～180℃之间（不要超过 185℃）[2]，至馏出液变清后，再继续反应 20min，尽可能将 1,2-二氯乙烷蒸出，然后冷却。待产品完全析出后抽滤，用少量水洗涤，干燥，称重。

粗产品加水，加热溶解，活性炭脱色，趁热过滤，冷却到 50～60℃时，将析出的白色柱状结晶过滤，得双酚-S，干燥，称重[3]。母液继续冷却到 20℃以下，析出双酚-S 的异构体，m.p.：185～190℃。

【附注】

[1] 也可加入苯等与水形成共沸物的溶剂，或减压将水蒸出。

[2] 成砜温度较高，一般在 160～180℃之间，温度过高，会有黑色焦油状聚合物产生。

[3] 双酚-S 与其异构体在 50～60℃时的溶解度相差较大，为了得到纯产品，有时需要进行多次重结晶。

Experiment 42 Synthesis of *p*-Toluene Sulfonic Acid

【Experimental Objectives】

(1) Understand the principle of sulfonation reaction.

(2) Learn and master the operation of water separation, recrystallization and drying of solid sample.

【Experimental Principles】

p-Toluenesulfonic acid (PTSA or pTsOH) or tosylic acid (TsOH) is an organic compound with the formula $CH_3C_6H_4SO_3H$. It is a white solid that is soluble in water, alcohols, and other polar organic solvents. As with other sulfonic acids, TsOH is a strong organic acid. It is about one million times stronger than benzoic acid. It is one of the few strong acids that is solid and, hence, conveniently weighed. Also, unlike some strong mineral acids (especially nitric acid, sulfuric acid, and perchloric acid), TsOH is non-oxidizing.

Toluene is easy to be sulfonated. The structure and proportion of sulfonation products are related to reaction temperature. Common impurities include benzenesulfonic acid and sulfuric acid. Impurities can be removed by recrystallization from its concentrated aqueous solution followed by azeotropic drying with toluene.

【Apparatus and Reagents】

(1) Apparatus: Round bottom flask, Dean-Stark trap (water separator), spherical condenser, Buchner funnel, suction flask, glass rod.

(2) Reagents: Toluene, concentrated sulfuric acid, hydrochloric acid.

【Experimental Procedure】

Add toluene (4mL) and concentrated sulfuric acid (1mL) into a round bottom flask provided with a Dean-Stark trap and a spherical condenser[1]. Heat the mixture gently to reflux until 0.3mL water is separated and afforded in the Dean-Stark trap. Stand a while to cool down the mixture, add several drops of water, then the crystals are precipitated. Stir with a glass rod to make the crystals precipitate completely. When filtrate, squeeze the mixture with a glass stopper to remove the toluene and *o*-toluene sulfonic acid, then the crude product is afforded and the yield is about 2g.

Crude products can be recrystallized by hydrochloric acid solution. Pure *p*-toluene sulfonic acid hydrate (p-$CH_3C_6H_4SO_3H \cdot H_2O$) is a colorless monoclinic crystal with a melting point of 104—105℃.

【Note】

[1] If the students can operate skillfully, the dosage of relative chemical reagents can be halved.

3.12 缩合反应

实验 43 双酚 A

【目的与要求】

（1）掌握酚醛（酮）缩合反应的机理。
（2）进一步熟悉重结晶等操作。

【原理】

双酚 A 是主要的有机合成中间体，是环氧树脂的重要原料。它通过苯酚与丙酮在催化剂硫酸和助催化剂作用下进行缩合反应而得到（如下所示）。双酚 A 又称：2,2-双（4,4′-二羟基二苯基）丙烷。

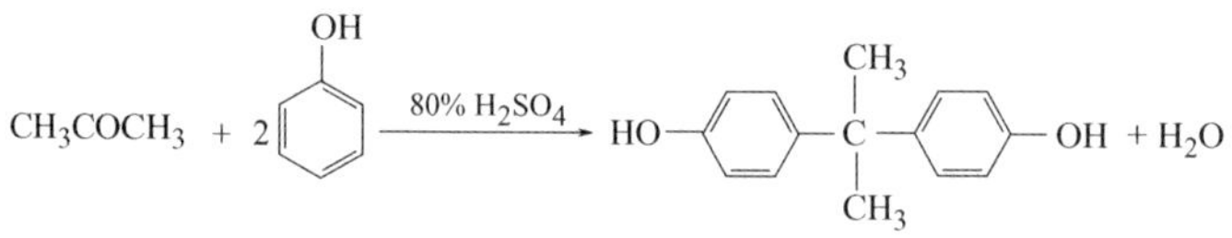

$$CH_3COCH_3 + 2 \underset{}{\bigcirc}\!-OH \xrightarrow{80\%\ H_2SO_4} HO-\underset{}{\bigcirc}\!-\underset{CH_3}{\overset{CH_3}{C}}-\underset{}{\bigcirc}\!-OH + H_2O$$

【仪器与药品】

（1）仪器：30mL 三口烧瓶，球形冷凝管，恒压滴液漏斗，砂芯漏斗，回流冷凝管。
（2）药品：丙酮，苯酚，80%硫酸，甲苯，硫代硫酸钠，氯乙酸。

【步骤】

在装有回流冷凝管和恒压滴液漏斗的 30mL 三口烧瓶中加入 0.5g $Na_2S_2O_3 \cdot H_2O$，加热熔化，再加入 0.2g 氯乙酸，混合均匀。然后依次加入 5g 苯酚、9mL 甲苯和 4mL 80%硫酸，滴加 2mL 丙酮，控制反应温度不超过 35℃。滴加完毕后，在 35～40℃下搅拌 2h。将产物倒入 25mL 冷水中，静置。完全冷却后过滤，冷水洗涤，得粗产物，产量约为 2.5g。粗产物可用甲苯重结晶提纯。

思 考 题

本实验中两分子苯酚与一分子丙酮在硫酸的催化作用下可生成哪些异构体？

实验 44 安 息 香

【目的与要求】

（1）了解安息香缩合反应的基本原理和维生素 B_1（盐酸硫胺素）为催化剂合成安息香

的方法。

(2) 掌握回流、重结晶等操作。

【原理】

以苯甲醛为原料,维生素 B_1(盐酸硫胺素)为催化剂,通过缩合反应制备目标产物安息香。反应式如下。

$$2\ PhCHO \xrightarrow[C_2H_5OH]{\text{维生素}B_1} PhCH(OH)COPh \qquad \text{维生素}B_1: \left[\begin{array}{c}\text{(噻唑盐结构)}\end{array}\right] Cl^- \cdot HCl$$

早期该反应的催化剂是剧毒的氰化物,后改用维生素 B_1,价格便宜,操作安全,效果良好。维生素 B_1 又叫硫胺素,是一种生物辅酶,其在生化过程 α-酮酸的脱羧和生成偶姻(α-羟基酮)等三种酶促反应中发挥辅助酶的作用。而在本实验中维生素 B_1 分子右边噻唑环上的 S 和 N 之间碳原子上的氢原子有较大的酸性,在碱的作用下生成碳负离子,催化苯偶姻的生成。反应机理如下:

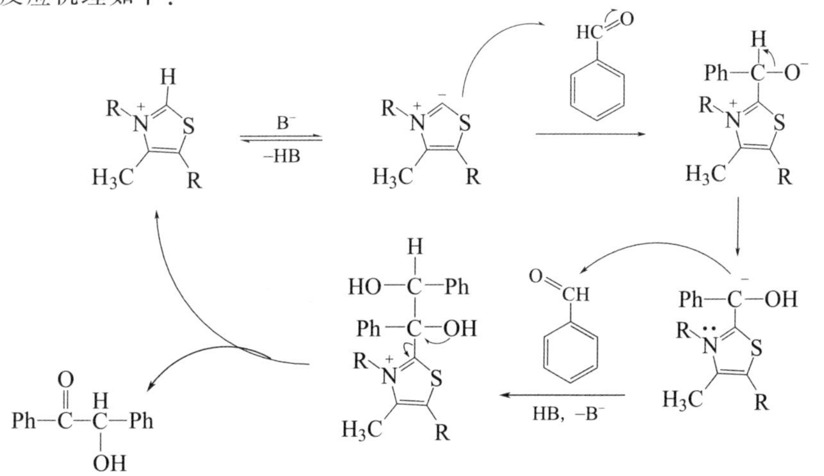

该反应的主要副反应为:

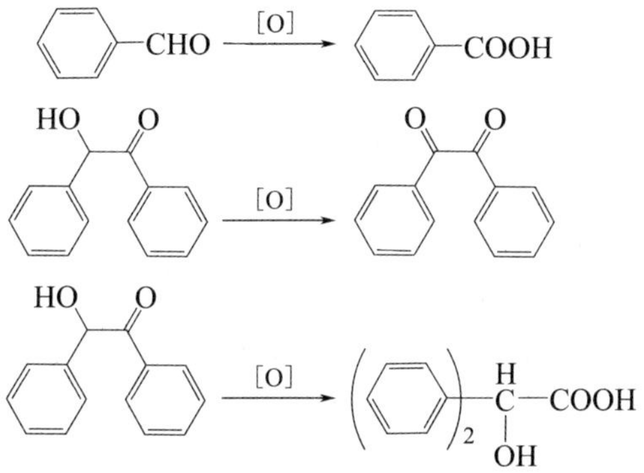

$$PhCHO \xrightarrow{[O]} PhCOOH$$

$$PhCH(OH)COPh \xrightarrow{[O]} PhCOCOPh$$

$$PhCH(OH)COPh \xrightarrow{[O]} Ph_2C(OH)COOH$$

【仪器与药品】

(1) 仪器:圆底烧瓶、电热套、球形冷凝管、滴液漏斗、分液漏斗、布氏漏斗、吸滤瓶。

(2) 药品:苯甲醛,维生素 B_1(盐酸硫胺素),95%乙醇。

【步骤】

在 25mL 圆底烧瓶中加入维生素 B_1 5.88g(0.025mol)、2mL 水和 95%乙醇 10mL。不时摇动,待维生素 B_1 溶解,将烧瓶在冰水浴中冷却,同时取 3mL 10%氢氧化钠放在试管中在冰水浴中冷却[1]。在冰水浴冷却下将 NaOH 加入烧瓶中，充分摇动，再加入新蒸馏的苯甲醛 5.25g（0.05mol），调节 pH＝9～10（必要时可加入氢氧化钠）[2]。等烧瓶冷却放在冰水浴中使之结晶析出，抽滤并用冷水洗涤（10mL×2），干燥，得到淡黄色的安息香粗品。粗品用 95％乙醇重结晶，必要时可加入活性炭脱色，产量 3g（产率 60％），测熔点。纯安息香为白色针状结晶，熔点为 137℃。

【附注】

[1] 维生素 B_1 在反应中作催化剂使用，它的质量对反应产生直接的影响。维生素 B_1 通常在酸性条件下稳定，易吸水，在水溶液中易被氧化而失效。同时光、金属离子（如铜、锰、铁等）均可加速维生素 B_1 的氧化，在氢氧化钠溶液中噻唑环易失效：

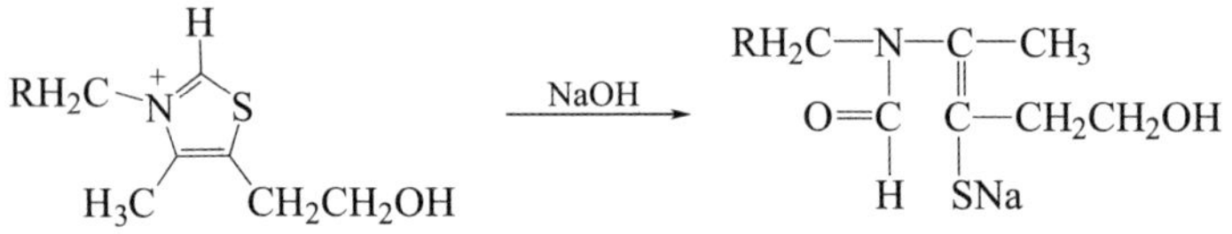

因此，反应第一步加入冰冷的氢氧化钠，其目的是为防止噻唑环发生开环反应，并促使维生素 B_1 形成负碳离子；维生素 B_1 溶液加碱时必须在冰浴冷却和搅拌下慢慢加入，加热时也不要过于剧烈。

[2] 加入反应原料后，反应体系必须保持溶液的 pH 为碱性。若溶液的 pH 值保持中性或偏酸性都不利于碳负离子的形成。

思 考 题

（1）写出苯甲醛在氰化钠作用下发生安息香缩合反应的反应机理。

（2）写出下述中间体的共振结构式，该中间体还有哪些催化性能（提示：卡宾）。

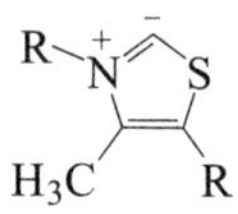

Experiment 45　Synthesis of Ethyl Acetoacetate

【Experimental Objectives】

(1) Learn the preparation principles and methods of operation of ethyl acetoacetate.

(2) Learn the anhydrous operation method.

【Experimental Principles】

When ethyl acetate reacts with sodium ethoxide, it undergoes a condensation reaction. After acidification, the product is a β-keto ester, ethyl acetoacetate (commonly called acetoacetic ester). Condensations of this type occur with many other esters and are known generally as Claisen condensations. Like the aldol condensation, Claisen condensations involve the α-carbon of one molecule and the carbonyl group of another.

In the Claisen condensation, dry ethyl acetate, containing a trace of absolute ethanol to which sodium is added, reacts to give ethyl acetoacetate too.

$$2CH_3CH_2OH \xrightleftharpoons[CH_3COOH]{NaOC_2H_5} CH_3COCH_2COCH_2CH_3 + C_2H_5OH$$

$$CH_3COCH_2CO_2C_2H_5 + NaOC_2H_5 \rightleftharpoons CH_3COCHCO_2C_2H_5^- Na^+ + HOC_2H_5$$

【Apparatus and Reagents】

(1) Apparatus: Heating mantle, three-necked round bottom flask, magnetic stirrer, spherical and straight condensate, separatory funnel.

(2) Reagents: Ethyl acetate, sodium, 5% acetic acid, saturated aqueous sodium chloride, anhydrous magnesium sulfate.

【Experimental Procedure】

Equip a 25mL three-necked round bottom flask with a mechanical stirrer, and with a reflux condenser. Attach a calcium chloride drying tube to the top of the condenser. Add 2mL of ethyl acetate[1] to the flask, and then 0.19g of clean finely sliced sodium is added as quickly as possible. The reaction starts immediately. Begin the stirrer, keep the reaction mixture in the flask slightly boiling (If it gets too violent, immerse the flask in a cold water bath). Keep the reaction mixture slightly boiling until the sodium has completely dissolved. This usually requires about one hour[2]. At this stage the reaction mixture should be a clean red liquid.

This solution is then cooled and made slightly acid by slowly adding 50% acetic acid (about 3mL) until the solution is just slightly acid to blue litmus paper[3]. If a solid appears, add a little of water to dissolve it. Pour the contents (not the boiling chip) into a separatory funnel, add an equal volume of saturated aqueous sodium chloride. Shake the mixture vigorously, and stand to separate into layer. The ester layer is separated and dried over anhydrous magnesium sulfate.

Filter through a funnel into a dry distilling flask. In order to minimize losses from ethyl acetoacetate clinging to the walls of the Erlenmeyer flask and the drying agent, rinse the flask with 2mL of ethyl acetate and pass the rinse through the funnel into the distilling flak. Remove unreacted ethyl acetate from the ethyl acetoacetate by distilling the filtrate on a boiling water bath. The residue is distilled under reduced pressure and the yield of ethyl acetoacetate on a boiling at 76—80℃/18mmHg is about 0.4g[4].

Ethyl acetoacetate is colorless. Literature: b. p. 181℃/760mmHg, 92℃/40mmHg, 82℃/20mmHg, 78℃/18mmHg, 74℃/14mmHg, 71℃/12mmHg.

【Notes】

[1] The ethyl acetate should be pure and anhydrous. Ordinary ethyl acetate can be purified following the procedure provided in appendix 4.

[2] It is usually asked that all the sodium metal disappear, but only thin unreacted slivers will not affect the following operation.

[3] It is advisable to avoid a large excess of acetic acid since it increases the solubility of the ester in water.

[4] During distillation, ethyl acetoacetate is easy to decompose at its atmospheric boiling point.

【Thinking Questions】

(1) In the preparation of ethyl acetoacetate, why is it essential that the reagents be free from water, and the apparatus perfectly clean and dry?

(2) Why is the solution red in color? When the sodium has completely dissolved.

(3) What are the purposes of the 50% acetic acid and saturated aqueous sodium chloride?

3.13 Diels-Alder 反应

实验46　9,10-二氢蒽-9,10-α,β-马来酸酐

【目的与要求】

掌握 Diels-Alder 反应的原理及其在有机合成上的应用。

【原理】

反应式：

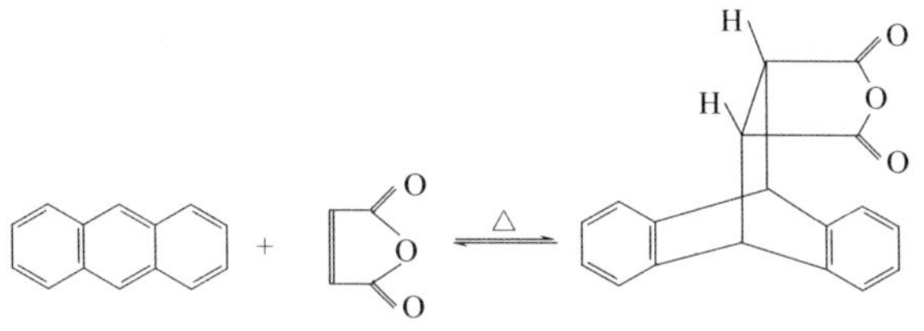

蒽（共轭双烯） + 马来酸酐（亲双烯试剂） $\xrightarrow{\triangle}$ 9,10-二氢蒽-9,10-α,β-马来酸酐（加成物）

【仪器与药品】

(1) 仪器：圆底烧瓶，球形冷凝管，三角漏斗，吸滤瓶，电热套，真空干燥器。
(2) 药品：蒽，无水二甲苯，马来酸酐。

【步骤】

准确称取 0.5g 马来酸酐和 1.0g 蒽加入 20mL 干燥的圆底烧瓶中，再加入 12mL 无水二甲苯[1]，放入两粒沸石，装上球形冷凝管，冷凝管上端加氯化钙干燥管。用电热套加热回流 30min[2]。反应结束后冷却至室温，过滤，用无水二甲苯重结晶粗产物，产量约为 0.5g。重结晶后，精制品放在含有石蜡屑的真空干燥器内干燥[3]后测其熔点。纯产品熔点为 263~264℃。

【附注】

[1] 马来酸酐和产物遇水均会水解成二元酸，因此反应仪器和所用试剂必须事先干燥。
[2] 延长加热回流时间可使产率提高。
[3] 干燥器内放入石蜡可以吸附二甲苯。如果干燥不好，产物吸收水分水解，对熔点测定有影响。

思　考　题

(1) Diels-Alder 反应的反应机理是什么？
(2) 本次实验中对水分有什么要求？最终产品为什么要放在含有石蜡屑的真空干燥器内干燥？

3.14 坎尼扎罗反应

实验47　呋喃甲醇和呋喃甲酸

【目的与要求】

了解呋喃甲醛的坎尼扎罗（Cannizzaro）反应原理和实验方法。

【原理】

无 α-H 的醛在强碱作用下，可发生自身的氧化还原反应，即一个分子醛还原为醇，另一个分子醛氧化为酸，这种反应称为坎尼扎罗反应。

本实验采用呋喃甲醛为反应物制备呋喃甲醇和呋喃甲酸。

反应式：

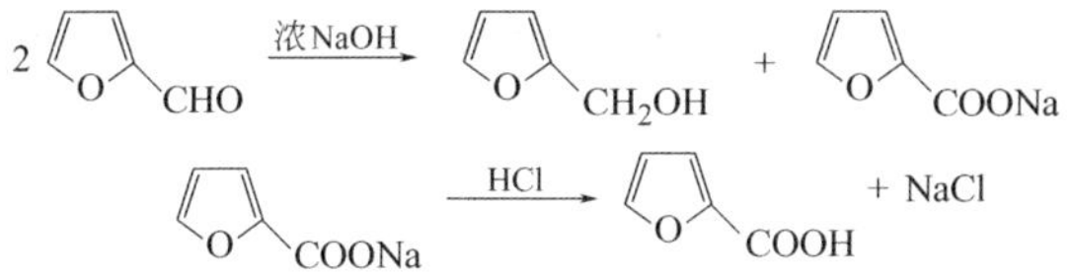

$$2 \text{ 呋喃-CHO} \xrightarrow{\text{浓 NaOH}} \text{ 呋喃-CH}_2\text{OH} + \text{ 呋喃-COONa}$$

$$\text{呋喃-COONa} \xrightarrow{\text{HCl}} \text{呋喃-COOH} + \text{NaCl}$$

【仪器与药品】

(1) 仪器：10mL 锥形瓶，分液漏斗，H 形分馏头，20mL 二口圆底烧瓶，三角滤板漏斗，5mL 和 10mL 茄形瓶，电磁加热搅拌器，滴管。

(2) 药品：呋喃甲醛，33％氢氧化钠溶液，浓盐酸，无水碳酸钠，活性炭，刚果红试纸，乙醚。

【步骤】

(1) 在10mL 锥形瓶内加入 3.2mL（1.86g，19.36mmol）呋喃甲醛和搅拌磁子，冰水浴冷却至 5℃，在电磁加热搅拌器的搅拌下，由滴管滴入 3.2mL 33％氢氧化钠溶液，控制温度在10℃左右[1]。加完后，于室温下搅拌 20min，成黄色浆状物。再加入约 1.5mL 水，使其恰好溶解，得棕红色溶液。

将棕红色溶液转移到分液漏斗中，用乙醚萃取 4 次，每次 1.5mL。合并有机层[2]，用适量无水碳酸钠干燥后，直接抽滤于二口圆底烧瓶中，装置如图 2-3(a) 进行蒸馏，先蒸出乙醚，再减压蒸出呋喃甲醇[3]，其产量为 0.5～0.7g，产率为 53％～73％。测定产品的折射率。纯呋喃甲醇 b.p. 为 171℃，n_D^{20} 为 1.4868。

(2) 乙醚萃取后的水溶液用约 2mL 25％盐酸（自配）酸化至刚果红试纸变蓝，充分冷却得浅黄色沉淀，抽滤，用少量冷水洗涤，压紧抽干。将初产品转移到10mL 锥形瓶中，加入 3.5mL 水，加热溶解后，稍冷，再加少许活性炭脱色，趁热用折叠滤纸过滤。滤液冷却，析出白色晶体，抽滤，烘干，得呋喃甲酸。其产量为 0.43～0.52g，产率为 45％～55％，实验产品熔点为 130～132℃。纯呋喃甲酸：m.p. 为 133～134℃。

【附注】

[1] 控温必须严格。高于 12℃，反应过于剧烈难以控制；若低于 8℃，反应速度太慢，会积累一些氢氧化钠，此后一旦反应，升温迅速，会增加副反应。

[2] 水层留待下一步实验用，不得丢弃。

[3] 呋喃甲醇的收集温度可根据系统的真空度由图 2-2 获得。

思 考 题

(1) 乙醚萃取后的水溶液用25％盐酸酸化到中性是否最合适？为什么？

(2) 本实验得两种产物，利用什么原理加以分离？

(3) 下列醛哪个能进行坎尼扎罗反应？写出反应产物和反应机理。

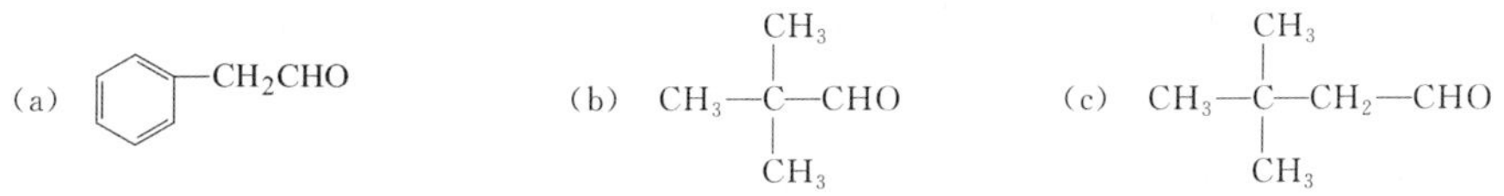

(a) $\text{C}_6\text{H}_5\text{-CH}_2\text{CHO}$ (b) $\text{CH}_3\text{-C(CH}_3)_2\text{-CHO}$ (c) $\text{CH}_3\text{-C(CH}_3)_2\text{-CH}_2\text{-CHO}$

(d) $CH_3-\underset{\underset{CH_3}{|}}{CH}-CHO$ (e) 4-methylbenzaldehyde (CHO and CH3 on benzene ring)   (f) $(CH_3)_2-\underset{\underset{CHO}{|}}{CH}-CH_2CH_3$

Experiment 48 Synthesis of Benzoic Alcohol and Benzoic Acid

【Experimental Objectives】

Study the principle and operation method of preparing benzoic alcohol and benzoic acid by Cannizzaro reaction.

【Experimental Principles】

The Cannizzaro reaction is the disproportionation reaction of the aldehyde class without α-active hydrogen under the action of strong NaOH or KOH water or alcohol solution. This reaction is characterized by the simultaneous oxidation and reduction of aldehyde itself, one molecule of aldehydes being oxidized to carboxylic acids (which become carboxylic acid salts in alkaline solutions), and the other a molecule of aldehydes being reduced to alcohols.

$$Ph-CH_2OH + NaOH \longrightarrow Ph-CH_2OH + Ph-COONa \xrightarrow{H^+} Ph-COOH$$

【Apparatus and Reagents】

(1) Apparatus: Heating mantle, round bottom flask, straight and spherical condensers, distillation head, H type fractionator, separatory funnel, conical flask, thermometer, Buchner funnel, suction flask, filter paper, watch glass, glass rod, measuring cylinder.

(2) Reagents: Benzaldehyde (fresh distilled), sodium hydroxide, diethyl ether, 10% sodium carbonate solution, concentrated hydrochloric acid, anhydrous magnesium sulfate, saturated sodium bisulfite solution, Congo red test paper.

【Experimental Procedure】

In a 25mL round-bottomed flask provided with a reflux condenser, mix thoroughly 0.9g sodium hydroxide and 4mL water. When sodium hydroxide dissolved, add 1.25mL fresh distilled benzaldehyde and a small amount of zeolite, heat the mixture to reflux for 1h. Continue heating the mixture until the benzoic acid has dissolved completely. The disappearance of oily benzaldehyde liquid indicates the completion of the reaction. Cool the contents of the flask slightly, remove the condenser, and set it downward for extraction[1].

(1) Preparation of benzyl alcohol

A sufficient amount of water (up to 10mL) is added to the reactants, which is constantly vibrating so that the benzoic acid salts are dissolved[2]. The solution is poured into the separatory funnel and extracted with diethyl ether (5mL×3). The upper ether extracts are washed with 5mL saturated sodium bisulfite solution, 16mL 10% sodium carbonate and 5mL water respectively. Then separate the upper layer of ether extract, dry with anhydrous magnesium sulfate. The dried ether solution is transferred to a 25mL round bottom flask, the common distillation unit is connected. Add zeolite and heat the mixture with warm water bath, then the ether is steamed out (recycled), and the 204-206℃ fraction is collected by direct heating. The boiling point of pure benzoic alcohol is 205.7℃

(2) Preparation of benzoic acid

After ether extraction, add concentrated hydrochloric acid to the water solution contained the sodium benzoate to make the Congo red test paper turn blue[3]. Fully stir and cool down to make benzoic acid precipitate completely. Then filter, recrystallize the crude product from water and measure its melting point. The melting point of pure benzoic acid is 122.4℃.

【Notes】

[1] When the liquid is separated by a separatory funnel, the water layer is divided from the bottom, and the ether layers are poured out from above, otherwise the operation will be affected.

[2] When drying with desiccant, must clarify before pour in distillation bottle, otherwise the product that steams out is impure.

[3] If the water layer is not completely acidic, the benzoic acid cannot be fully precipitated, resulting in loss of the product.

【Thinking Questions】

(1) The Cannizzaro reaction occurs much more slowly in dilute than in concentrated potassium hydroxide solution. Why?

(2) What does the washing of each department remove?

(3) Is the aqueous solution of extraction the most suitable for acidification to neutral? Why? Without the test paper, how to know the acidification is appropriate.

3.15 格利雅反应

实验49 苯 甲 酸

【目的与要求】

(1) 掌握格氏试剂的制备方法。

(2) 掌握格氏反应制备羧酸的方法。

【原理】

主反应:

$$\text{C}_6\text{H}_5\text{-Br} + \text{Mg} \xrightarrow{\text{Et}_2\text{O}} \text{C}_6\text{H}_5\text{-MgBr}$$

$$\text{C}_6\text{H}_5\text{-MgBr} + \text{CO}_2 \longrightarrow \text{C}_6\text{H}_5\text{-C(=O)-OMgBr} \xrightarrow{\text{H}^+} \text{C}_6\text{H}_5\text{-C(=O)-OH}$$

副反应：C₆H₅MgBr + C₆H₅Br ⟶ C₆H₅—C₆H₅ + MgBr₂

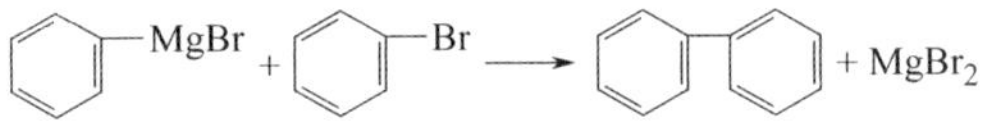

【仪器与药品】

(1) 仪器：二口圆底烧瓶，滴液漏斗，分液漏斗，干燥管，三角漏斗，吸滤瓶，冷凝管。

(2) 药品：溴苯，镁，无水乙醚，干冰，碘，浓盐酸，5%氢氧化钠溶液，活性炭。

【步骤】

本实验第一步反应所用的药品必须是无水的，所用的仪器必须是干燥的。

(1) 苯基溴化镁的制备：在干燥的二口圆底烧瓶上装好滴液漏斗和冷凝管。滴液漏斗和冷凝管上口都装上氯化钙干燥管以防止空气中的湿气进入。向烧瓶中加入 0.5g 干燥的镁屑[1]、6mL 无水乙醚[2]、2.2mL 溴苯和 1 小粒碘。如果反应在几分钟内仍未发生[3]，可将烧瓶温热。若反应过于激烈可适当用冷水控制。15min 后加热回流并搅拌反应，直至镁全部反应完毕形成灰白色黏稠液体。冷却后立即用此溶液做下面的实验。

(2) 苯甲酸的制备：用冰盐浴将上述反应混合物冷却到 −5℃。从瓶口插入导气管，通入由干冰产生的二氧化碳气体 25min[4]。开始通入的二氧化碳气体很快被吸收，到反应完毕时，反应器中二氧化碳压力上升[5]，即可切断二氧化碳气流。此时瓶内形成灰色黏稠固体。在冰盐浴冷却下，缓慢滴加由冰水与浓盐酸配成的 1∶1 混合液。开始滴加时速度要慢，以免生成的二氧化碳将产物冲出，并不时搅拌到固体溶解（如果物质过于黏稠可适当加入少量乙醚）。

将反应液转移到分液漏斗中，分出有机相，将水相用 8mL 乙醚分 2 次萃取，合并醚层有机相。醚层用 8mL5%氢氧化钠溶液分 2 次萃取，取碱性水溶液放在烧杯内，加热以除去溶在水中的乙醚[6]（若溶液的颜色较深可用活性炭脱色）。冷却，边搅拌边缓慢滴加浓盐酸直到溶液显酸性。在冰水中搅拌冷却，充分析晶后抽滤得粗品，产量约为 1g。粗品用水重结晶后测定熔点。纯苯甲酸熔点为 122.4℃。

【附注】

[1] 可用镁条代替镁屑。镁条在使用前用细砂纸将其表面擦亮（除去氧化层），并剪成小段。

[2] 市售乙醚使用前要做除水处理。将乙醚先用无水氯化钙浸泡 24h，间歇振摇以除去大部分水和乙醇。蒸馏，收集 34~35℃馏分。在收集瓶中压入钠丝直到钠丝上无气泡产生为止，密闭放置 24h 以上，待乙醚中残留的痕量水和乙醇转化为氢氧化钠和乙醇钠后方能使用。

[3] 反应开始时溶液先变白，后又变为棕色，并逐渐加深。反应产生的热量可使乙醚沸腾。

[4] 二氧化碳气体可用大理石加盐酸自制或从钢瓶中获得。

[5] 将回流冷凝管上的氯化钙干燥管出口用橡皮管与一个吸滤瓶的支管相连。吸滤瓶中盛 1/3 水，瓶口插单孔塞，上插一根玻璃管，管的下端升到水面下。若反应器中压力升高，则吸滤瓶玻璃管中的液柱升高。当液柱上升到 20~30cm 时即可停止通二氧化碳。

[6] 如果乙醚不除尽，制得的苯甲酸可能为蜡状固体而得不到晶体。

思 考 题

(1) 为什么制成的苯基溴化镁溶液应立即用来做下一步加成反应？

(2) 如何控制副产物联苯的产生？

(3) 干冰含有冷凝水会产生什么副反应？

实验 50　2-甲基-2-己醇

【目的与要求】

进一步掌握格氏反应的原理和反应操作。

【原理】

反应式：

$$n\text{-}C_4H_9Br + Mg \xrightarrow{\text{无水乙醚}} n\text{-}C_4H_9MgBr$$

$$n\text{-}C_4H_9MgBr + H_3CCCH_3 \xrightarrow{\text{无水乙醚}} n\text{-}C_4H_9\underset{\underset{OMgBr}{|}}{C}(CH_3)_2$$
$$\quad\quad\quad\quad\quad\quad\quad\underset{O}{\|}$$

$$n\text{-}C_4H_9\underset{\underset{OMgBr}{|}}{C}(CH_3)_2 + HOH \xrightarrow{H^+} n\text{-}C_4H_9-\underset{\underset{OH}{|}}{C}(CH_3)_2$$

【仪器与药品】

(1) 仪器：二口圆底烧瓶，电磁加热搅拌器，恒压滴液漏斗，冷凝管，干燥管，分液漏斗。

(2) 药品：丙酮，无水乙醚，10%硫酸，5%碳酸钠溶液，无水碳酸钾，无水氯化钙，镁，碘，正溴丁烷。

【步骤】

(1) 正丁基溴化镁的制备：在二口圆底烧瓶上分别装好冷凝管及恒压滴液漏斗，在冷凝管上口装上氯化钙干燥管。烧瓶内加入 0.9g 镁屑（或除去氧化膜的镁条）、4.5mL 无水乙醚[1]及 1 小粒碘。在恒压滴液漏斗中混合 3.9mL 正溴丁烷和 4.5mL 无水乙醚。先向瓶内滴入约 2mL 混合液，数分钟后溶液呈微沸状态，碘的颜色消失。若不发生反应可用温水浴加热。若反应比较剧烈，可适当用冷水浴冷却。待反应缓和后，自冷凝管上端加入 6mL 无水乙醚。开动电磁加热搅拌器搅拌，并滴入剩余的正溴丁烷-乙醚混合液。控制滴加速度维持反应液呈微沸状态。滴加完毕后，在水浴上回流 10min，使镁屑作用完全。

(2) 2-甲基-2-己醇的制备：将上述制好的格氏试剂在冰浴冷却和搅拌下，自滴液漏斗滴入 2.8mL 丙酮和 4.5mL 无水乙醚的混合液，控制滴加速度勿使反应过于猛烈。加完后，室温搅拌反应 10min。溶液中有白色黏稠状固体析出。

将反应瓶在冰浴冷却和搅拌下，自滴液漏斗滴加 30mL 10%硫酸溶液分解产物。待分解完全后，将反应混合物转移至分液漏斗中，分出醚层。水层每次用 8mL 乙醚萃取 2 次，将醚层合并，用 8mL 5%碳酸钠溶液洗涤后，用无水碳酸钾干燥[2]。干燥后的粗产物醚溶液滤入蒸馏瓶，用温水浴蒸去乙醚，再蒸馏出产品，收集 137～141℃ 馏分，其产量约为 1.4g。

纯 2-甲基-2-己醇的沸点为 143℃。

【附注】

[1] 镁条及乙醚均要事先做无水处理。方法见上一个实验。

[2] 2-甲基-2-己醇与水能形成共沸物。所以干燥必须完全，否则前馏分将大大增加。

思　考　题

(1) 格氏反应为什么要严格控制水分？如何控制？

(2) 为什么本实验制得的粗产品不能用无水氯化钙干燥？

(3) 2-甲基-2-己醇还可以用什么方法合成？

Experiment 51　Synthesis of Malachite Green and Crystal Violet

【Experimental Objectives】

(1) Learn the principle of Grignard reaction.

(2) Learn and master the operations of Schlenk line tube.

【Experimental Principles】

Grignard reagents are prepared from organic halides by treatment with magnesium metal. The reagents are sensitive to both oxygen in air and water, so they are prepared under a protective atmosphere of nitrogen and reacted directly with the electrophile.

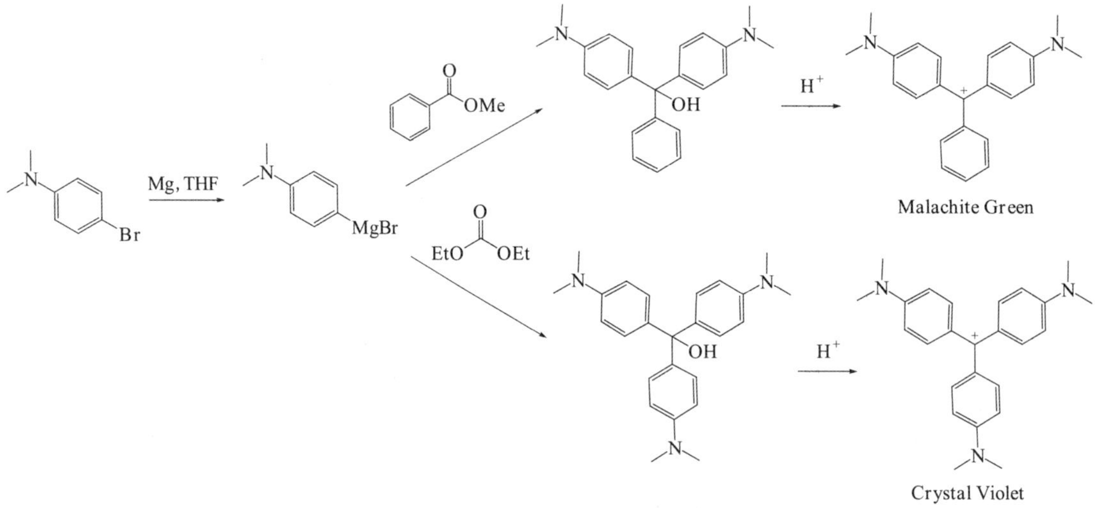

Grignard reactions are almost always done in either diethyl ether or THF. The oxygen atoms of these solvents can stabilize the reagent by donation of lone pair electrons to magnesium. The solvents must be specially dried and deoxygenated before use.

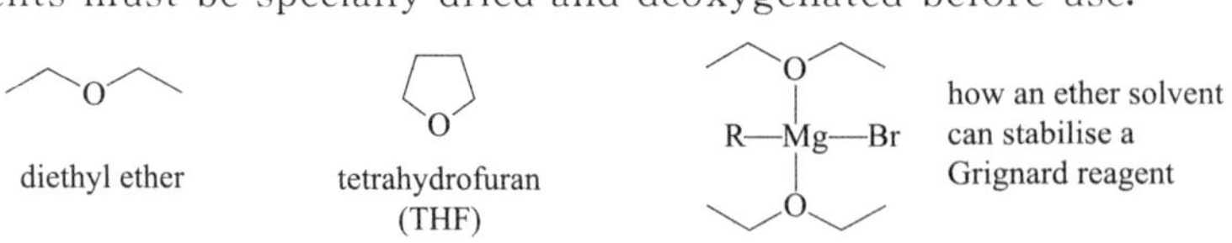

In this experiment, the organic halide will be 4-bromo-N,N-dimethylaniline. The Grignard reagent will be reacted with either methyl benzoate or diethyl carbonate. Interestingly, the products of these reactions, as tertiary alcohols with three aryl groups, readily ionise on addition of acids to give highly resonance stabilized carbocations. These cations are strongly colored and can be used as dyes[1].

【Apparatus and Reagents】

(1) Apparatus: Round bottom flask, heating mantle spherical condenser, glass funnel, cotton cloth.

(2) Reagents: Magnesium metal, 4-bromo-N,N-dimethylaniline, anhydrous THF, hydrochloric acid, iodine.

【Experimental Procedure】
From the oven, take a two necked 25mL round bottom flask and a reflux condenser. Clamp the flask securely above a hotplate stirrer. Fit the condenser into the center neck of the flask and fit the adapter of the Schlenk line tube into the top of the condenser. Seal the second neck of the flask with a septum. Turn the tap of the Schlenk line to vacuum and wait until the glassware has cooled to room temperature.

Weigh out magnesium metal (100mg) and 4-bromo-N,N-dimethylaniline (0.625g). Turn the tap of the Schlenk line to nitrogen to release the vacuum inside the flask, then turn the tap off. Add the magnesium and 4-bromo-N,N-dimethylaniline through one of the necks. Turn the Schlenk line tap back to nitrogen. Using a 20mL syringe, collect anhydrous THF (5mL) from the flask prepared by the lab staff. Inject it into your flask through the septum. Quickly drop in a small crystal of iodine by briefly lifting up the condenser. Connect the lower connector of the condenser to the water supply and the outer connecter to the return tube. The flow of cooling water will ensure that no THF is lost. Using a heating mantle, heat the reaction mixture at gentle reflux for thirty minutes. At the end of this time, lift the flask out of the heating mantle and let it cool to room temperature. Using a smaller syringe (CAUTION-the needles are sharp), inject the volume of either diethyl carbonate or methyl benzoate that you calculated in the pre lab exercise. Heat at reflux for another 5min, then cool to room temperature again. Turn off the cooling water and remove the condenser. Cautiously add 10% hydrochloric acid (2mL) bit by bit, since the reaction with any left-over magnesium metal will be vigorous with rapid evolution of H_2.

Pour the muddy mixture into a beaker. Immerse the piece of cotton cloth for one minute, then take it out and rinse with water. Do the rinsing in the fume cupboard in a bucket or bowl (to be supplied by the lab staff). Do not rinse in the sink. After the cloth has been hung up to dry, put it in a plastic bag, label it with name or numbers.

【Notes】
[1] The dyes easily stain hands, clothing, furniture etc. Wear gloves and take care. Both dyes are toxic. Do not use the dyes on clothing or any personal articles.

3.16 催化氢化反应

实验52 3-苯基丙酸（氢化肉桂酸）

【目的与要求】
(1) 学习 Raney 镍氢化催化剂的制备方法。
(2) 掌握微型催化加氢反应的操作。

【原理】
反应式：
$$NiAl_2 + 2NaOH + 2H_2O \longrightarrow Ni + 2NaAlO_2 + 3H_2\uparrow$$
$$C_6H_5CH=CHCO_2H + H_2 \xrightarrow[\text{常温常压}]{\text{Ni}} C_6H_5CH_2CH_2CO_2H$$

【仪器与药品】

(1) 仪器：常压催化氢化装置，圆底烧瓶，分液漏斗，氢气袋，电磁搅拌器。

(2) 药品：镍铝合金，肉桂酸，氢氧化钠，无水乙醇。

【步骤】

(1) Raney 镍的制备：在烧杯中加入 2g 镍铝合金（含 Ni40%～50%）及 10mL 蒸馏水，搅拌均匀。然后分批加入 3.2g 氢氧化钠，并加以搅拌。反应放热，并有氢气逸出。控制氢氧化钠的加入速度，以泡沫不溢出为宜，至无明显的气泡逸出为止（约需 5min）。反应物在室温放置 10min，然后在 70℃ 水浴中保温 0.5h，倾去上层清液，用蒸馏水以倾析法洗 2 次，再用乙醇洗 2 次，用无水乙醇覆盖备用[1]。

(2) 肉桂酸的催化氢化：简易常压催化氢化装置如图 3-10 所示，三通活塞 1 接氢气囊，三通活塞 2 接真空系统。

在圆底烧瓶中溶解 0.1g 肉桂酸于 2mL 温热的无水乙醇中，冷至室温。加入 4～8 滴的 Raney 镍乙醇，用少量乙醇冲洗瓶壁。放入搅拌磁子后塞紧有导气管的橡皮塞，与氢化系统相连。检查整个系统是否漏气[2]。

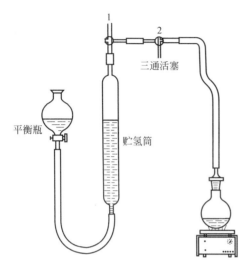

图 3-10 常压催化氢化装置图

氢化开始前，调节图 3-10 中活塞 1 通向大气，把盛有蒸馏水的平衡瓶的位置提高，使贮氢筒内充满水，赶尽瓶内的空气。关闭活塞 1，打开三通活塞 2，使与真空系统相连，抽真空排除整个氢化系统内的空气。抽到一定真空度后关闭活塞 2，打开与氢气袋相连的活塞 1 进行充氢。如此抽真空-充氢，重复 2～3 次，即可排除整个系统内的空气[3]。最后再关闭与真空系统相连的活塞 2，打开活塞 1，使氢气与贮氢筒连通，同时降低平衡瓶的位置，使贮氢筒内充满氢气，关闭活塞 1，断开与氢气袋的连接。

取下平衡瓶，使其平面与贮氢筒中水平面高度保持水平，记下贮氢筒内氢气的体积，即可开始氢化反应。开动搅拌，隔一定时间记录吸氢体积（量体积时放下平衡瓶使与贮氢瓶水面相平），作出时间-吸氢体积曲线。当贮氢瓶内氢气体积不再有明显下降时，说明氢化反应已经完成。关闭三通活塞 1，断开与反应体系的连接，打开三通活塞 2 放掉氢化瓶中的残余氢气。取下氢化反应器，抽滤除去催化剂。催化剂应放入专门的回收瓶中，切忌随便丢入废物缸内，以免引起着火事故。滤液先在水浴上蒸去乙醇，趁热倒在表面皿上，冷却后得到氢化肉桂酸结晶。干燥后称重。如需进一步纯化，可减压蒸馏，收集 145～147℃/2.4kPa 的馏分。

【附注】

[1] 该法制备的催化剂是略带碱性的高活性催化剂。催化剂的贮存会导致活性显著降低。因此最好新鲜制备，可得到较高转化率。催化剂制好后，挑少许于滤纸上，待溶剂挥发后，催化剂能发火自燃，表示活性较好，否则需要新制备。

[2] 检查的方法是，将整个氢化系统与带有压力计的水泵相连。开启水泵，当抽至一定压力后，关闭水泵。切断与氢气系统的连接，观察压力计的读数是否发生变化。若系统漏

气，应逐次检查玻璃活塞、磨口塞是否塞紧及橡皮管连接处是否紧密等。

[3] 氢化反应前必须排除系统内的空气，氢化过程中严禁空气进入氢化系统内。

思 考 题

(1) 氢化反应过程应注意哪些事项？

(2) 为什么氢化过程中要搅拌？

Experiment 53　Synthesis of *cis*-4-*t*-Butylcyclohexanol

【Experimental Objectives】

(1) Understand the stereoselective reduction.

(2) Master the operation of high pressure reactor.

【Experimental Principles】

Aromatic compound can be stereoselective hydrogenated with the presence of Raney nickel.

$$HO-C_6H_4-C(CH_3)_3 + H_2 \xrightarrow{\text{Raney Ni}} HO-C_6H_{10}-C(CH_3)_3$$

cis：*trans* = 3：1

【Apparatus and Reagents】

(1) Apparatus: High pressure reactor.

(2) Reagents: Raney nickel, isopropanol, 4-*t*-butylphenol, hydrogen.

【Experimental Procedure】

In the 100mL high pressure reactor[1], 15g 4-*t*-butylphenol, 1g Raney nickel, and 30g isopropanol are added. The reaction system should be sealed, and pumped in vacuum and filled the nitrogen cycle three times. Then fill the hydrogen in the sealed reaction system. This system is pressured to 5atm and heated to 130℃ for keeping 2h. After cooling the reaction to room temperature and normal pressure, the mixture should be filtered and removed the solvent to afford the crude product. The ratio between *cis*- and *trans*-4-*t*-butyl-cyclohexanol is 3：1.

【Notes】

[1] High-pressure hydrogenation reaction is very dangerous! Please according to the instruction to use the high pressure reactor carefully.

Experiment 54　Reduction of Camphor

【Experimental Objectives】

(1) Learn the principle and operation method of reducing camphor with $NaBH_4$.

(2) Understand the application of thin layer chromatography in synthesis reaction.

【Experimental Principles】

Borneol is a bicyclic organic compound and a terpene derivative. The hydroxyl group in this compound is placed in an endo position. There are two different enantiomers of borneol. Both d-(+)-borneol and l-(−)-borneol are found in nature. Whereas d-borneol was the enantiomer that used to be the most readily available commercially. The more commercially available enantiomer now is l-borneol, which also occurs in nature. Borneol is used in traditional Chinese medicine. An early description is found in the Bencao Gangmu. Borneol is a component of many essential oils and it is a natural insect repellent. Derivatives of isoborneol are used as chiral ligands in asymmetric synthesis.

Borneol can be synthesized by reduction of camphor by the Meerwein-Ponndorf-Verley reduction (a reversible process). Reduction of camphor with sodium borohydride (fast and irreversible) gives instead the isomer isoborneol as the kinetically controlled reaction product.

【Apparatus and Reagents】

(1) Apparatus: Conical bottles, glass rods, Brinell funnel, thin laminates, tomographic cylinders.

(2) Reagents: Camphor, sodium borohydride[1], methanol, ether[2], anhydrous sodium or magnesium sulfate.

【Experimental Procedure】

To the camphor (1.0g) in the 25mL Erlenmeyer flask, add methanol (10mL). Stir with a glass rod or microspatula until the solid has dissolved. In portions, cautiously and intermittently add 0.6g of sodium borohydride to the solution at room temperature. If necessary, cool the Erlenmeyer flask in an ice bath to control the temperature of the reaction mixture. When all the borohydride is added, heat the contents of the flask to reflux until all the sodium borohydride has dissolved.

Allow the reaction mixture to cool to room temperature, and then carefully add to ice water (20mL) while stirring. Collect the white solid which forms by filtering through a Buchner funnel, and suction-dry it for several minutes while you clean and dry the 100mL Erlenmeyer flask. Transfer the solid back to the clean flask, and add 25mL of ether to dissolve the solid followed by 6—7 microspatulafuls of anhydrous sodium or magnesium sulfate. After about 5min of drying, transfer the solution (leaving behind the drying agent) to a previously weighed beaker or Erlenmeyer flask. Evaporate the solvent off in the hood to obtain the product as a white solid. Recrystallize the crude product from aqueous ethanol. The yield of the product is about 0.9g.

Weigh the product and calculate the percentage yield. Determine the melting point (literature melting point: isoborneol 212℃) in a sealed capillary tube and record the IR spectrum a KBr pellet.

【Notes】

[1] Sodium borohydride liberates hydrogen upon reaction with water, so it should be kept it in a drying condition.

[2] Ether is highly flammable, so keep it away from fire.

3.17 天然产物的提取和分离

实验 55 绿色植物色素的提取

【目的与要求】
(1) 学习绿色植物色素的提取原理和提取方法。
(2) 了解色谱分离的原理。
(3) 掌握柱色谱的操作技能。

【原理】
绿色植物的茎、叶中含有叶绿素（绿）、胡萝卜素（橙）和叶黄素（黄）等多种天然色素。叶绿素有两个异构体，即叶绿素 a（$C_{55}H_{72}O_5N_4Mg$）和叶绿素 b（$C_{55}H_{70}O_6N_4Mg$），其差别仅是叶绿素 a 中一个甲基被甲酰基所取代而形成了叶绿素 b。它们都是吡咯衍生物与金属镁的络合物，是植物进行光合作用所必需的催化剂。植物中叶绿素 a 的含量通常是叶绿素 b 的 3 倍。尽管叶绿素分子中含有一些极性基团，但大的烃基结构使它易溶于醚、石油醚等一些非极性的溶剂。胡萝卜素（$C_{40}H_{56}$）是具有长链结构的共轭多烯，它有三种异构体，即 α-胡萝卜素、β-胡萝卜素和 γ-胡萝卜素，其中 β-胡萝卜素最多，也最重要。在生物体内，β-胡萝卜素受酶催化氧化形成维生素 A。叶黄素（$C_{40}H_{56}O_2$）是胡萝卜素的羟基衍生物，它在绿叶中的含量通常是胡萝卜素的两倍。与胡萝卜素相比，叶黄素较易溶于醇，而在石油醚中的溶解度较小。

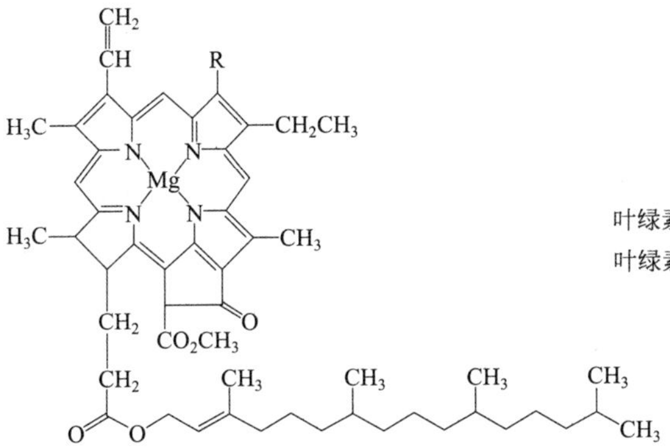

叶绿素a (R=CH$_3$)
叶绿素b (R=CHO)

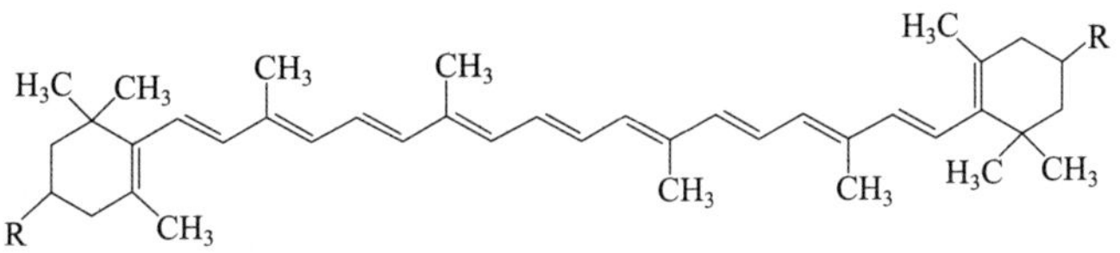

β-胡萝卜素 (R=H)
叶黄素 (R=OH)

【仪器与药品】
(1) 仪器：研钵，布氏漏斗，抽滤瓶，烧杯，展开缸（或广口瓶），薄层板（硅胶 G），色谱柱。
(2) 药品：绿色植物叶，石油醚（60～90℃），丙酮，乙醇，乙酸乙酯，正丁醇，毛细管，硅胶（或中性三氧化二铝）。

【步骤】

(1) 绿色植物色素提取

将5g绿色植物叶子用剪刀剪碎并与5mL乙醇拌匀,置于研钵中研磨约5min,用30mL石油醚-乙醇混合液(2:1,体积比)分次浸取。深绿色浸取液过滤、合并,转入分液漏斗中,加入等体积的水洗涤,以除去乙醇以及浸取液中的少量水溶性物质。洗涤时注意不要剧烈振荡,以防止乳化。静置分层后从分液漏斗下方放出水层(水-乙醇层),上方有机层(石油醚层)转移到烧杯中,加入少量无水硫酸钠干燥、浓缩、备用。

(2) 薄层色谱

用毛细管吸取绿色植物色素的提取液,在薄层板起点线上进行点样,并使点尽可能小而均匀。(待溶剂完全挥发后,再吸取提取液重新点在同一位置上进行点样,反复点样5~8次)。分别用石油醚-丙酮(8:2,体积比)和石油醚-乙酸乙酯(6:4,体积比)两种溶剂系统作为展开剂,将点样后的薄层板,小心放入预先加入选定展开剂的展开缸(或广口瓶)内,盖好瓶盖。待展开剂上升至规定高度时,取出薄层板,在空气中晾干,观察,从上到下四个斑点依次为:胡萝卜素(橙色)、叶黄素(黄色)、叶绿素a(蓝绿色)叶绿素b(黄绿色)。用铅笔进行标记,并分别计算出各分离物质的比移值(R_f值)。注意更换展开剂时,须干燥展开缸(或广口瓶)。

(3) 柱色谱

安装好色谱柱,取柱色谱硅胶进行湿法装柱。在色谱柱中,加约3cm高的石油醚,另取少量脱脂棉,用石油醚浸湿、挤压、驱除气泡后,置于色谱柱底部,轻轻压紧,塞住底部。将柱色谱硅胶与石油醚搅拌成糊状,并将其慢慢倒入色谱柱中,同时打开底部活塞,让石油醚流入接收瓶中,必要时用橡胶棒(或洗耳球)轻轻敲打色谱柱,使吸附剂装得均匀致密,没有裂缝和气泡。装完后,在柱内吸附剂上面加0.5cm厚的石英砂或一片圆形滤纸,并不断用溶剂石油醚洗脱,以使色谱柱流实。然后放掉过剩的溶剂,直至溶剂面刚刚达到石英砂或滤纸的顶部,关闭活塞(注意!在任何情况下,硅胶表面不得露出液面)。

用胶头滴管吸取适量上述绿色植物色素提取液,小心地加到色谱柱顶部(约1cm高)。加完后,打开下端活塞,让液面下降到柱面以上1mm左右,关闭活塞,加数滴石油醚,打开活塞,使液面下降,经几次反复,使色素全部进入柱体。待色素全部进入柱体后,在柱顶小心加入洗脱剂进行洗脱。分别用石油醚-丙酮(9:1,体积比)、石油醚-丙酮(7:3,体积比)和正丁醇-乙醇-水(3:1:1,体积比)为洗脱剂进行洗脱,依次接收各色素带,即得胡萝卜素(橙黄色溶液)、叶黄素(黄色溶液)、叶绿素a(蓝绿色溶液)以及叶绿素b(黄绿色溶液)。

【注释】

叶黄素易溶于醇而在石油醚中溶解度较小,从绿色植物(如绿菠菜)得到的提取液中,叶黄素含量很少,柱色谱中不易分出黄色带。

【思考题】

试比较叶绿素、叶黄素和胡萝卜素三种色素的极性,为什么胡萝卜素在色谱柱中移动最快?

实验56 从西红柿中提取番茄红素和 β-胡萝卜素

【目的与要求】

(1) 熟悉从天然植物中分离色素的原理和方法。

(2) 掌握柱色谱的操作技能。

【原理】

食品的色泽是构成食品感官质量的一个重要因素，保持或赋予食品良好的色泽的方法之一就是添加色素。色素分人工合成色素和天然色素两大类。一般合成色素都有一定的毒性，因此人们倾向使用天然色素。天然色素从来源上可分为植物色素、动物色素和微生物色素，其中植物色素缤纷多彩，是常用的色素。

β-胡萝卜素和番茄红素都属于类胡萝卜素，其分子式为 $C_{40}H_{56}$，结构式如下。

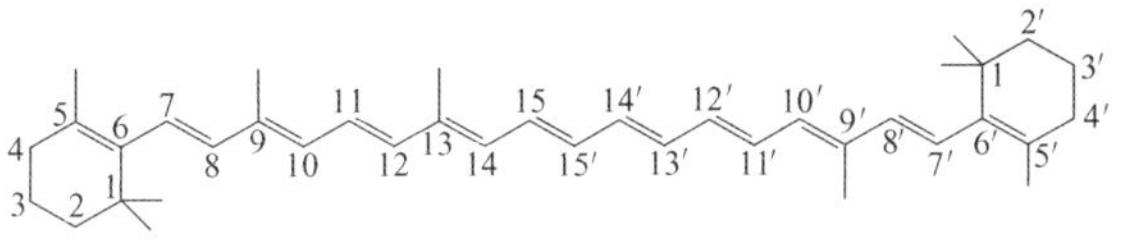

β-胡萝卜素

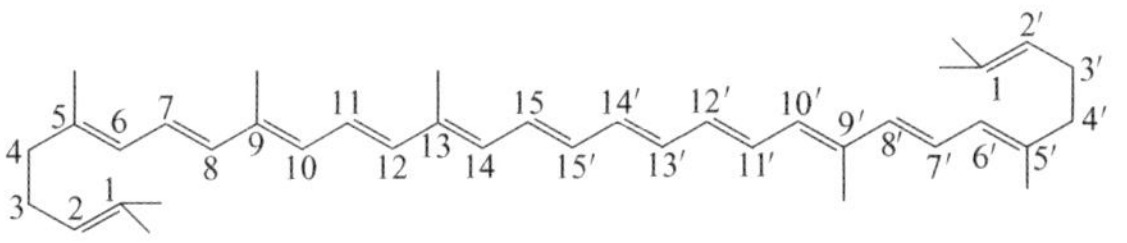

番茄红素

由于 β-胡萝卜素从中间断裂可形成二分子的维生素 A，因此是一种廉价的维生素 A，既是天然色素，又是营养强化剂。

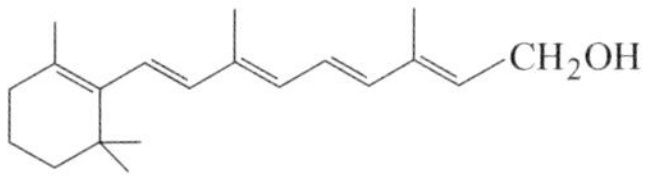

维生素 A

类胡萝卜素属于脂溶性物质。在实验中采用二氯甲烷作为萃取剂，由于二氯甲烷与水不能混溶，因此先用乙醇除去番茄中的水。提取出的粗产品用柱色谱分离。

【仪器与药品】

(1) 仪器：25mL 圆底烧瓶，球形冷凝管，抽滤装置，分液漏斗，色谱柱，滴液漏斗。

(2) 药品：西红柿，95% 乙醇，二氯甲烷，Al_2O_3（薄层），苯，环己烷，石油醚，氯仿，氯化钠，无水硫酸钠。

【步骤】

(1) 提取

将 10g 新鲜西红柿捣碎，加入到 25mL 的圆底烧瓶中，再加入 15mL 95% 乙醇，摇匀，装上回流冷凝管。加热回流 10min，趁热倒出上层溶液，再加入 10mL 二氯甲烷，回流 5min，冷却，将上层溶液倒出，再加 10mL 二氯甲烷重复萃取 1 次。合并乙醇和二氯甲烷萃取液，过滤，将滤液加入分液漏斗中，加几毫升氯化钠溶液，振摇，静置分层。将分出的二氯甲烷溶液用无水硫酸钠干燥，蒸馏除去二氯甲烷，备用。

(2) 色谱柱的装填

将 $10gAl_2O_3$ 与 10mL 苯搅拌成糊状，并将其慢慢加入预先加了一定苯的色谱柱中，同时打开活塞，让苯流入接收瓶中，不时用橡胶棒敲打色谱柱，以稳定的速度装柱，使色谱柱装得均匀。装好的柱子不能有裂缝和气泡。在装好的柱子上放 0.5cm 厚的石英砂或一片滤纸，并不断用溶剂苯洗脱，以使色谱柱流实。然后放掉过剩的溶剂，直至溶剂面刚刚达到石英砂或滤纸的顶部，关闭活塞。

(3) 洗脱

将粗类胡萝卜素溶解于尽量少的苯中,用滴管加入柱顶,打开活塞,让溶剂滴下,待溶剂面刚刚达到石英砂或滤纸的顶部,再用滴管加入几毫升苯,然后用环己烷/石油醚(1∶1)30mL洗脱,黄色的 β-胡萝卜素在柱中的流动较快,红色的番茄红素移动较慢。收集洗脱液至黄色的 β-胡萝卜素从柱上完全除去。最后换接收瓶,用氯仿作洗脱剂,洗脱番茄红素。将收集到的两种洗脱液分别蒸馏至干。

【备注】

β-胡萝卜素对光及氧非常敏感,对酸、碱也敏感,重金属离子特别是铁离子也可使其颜色消失。番茄红素的耐光、耐氧化性也很差。所以在实验中要特别注意。

实验57 果胶的提取

【目的与要求】

(1) 了解果胶的性质和提取原理。
(2) 掌握果胶的提取工艺。
(3) 了解果胶在食品工业中的用途。

【原理】

果胶广泛存在于水果和蔬菜中。例如苹果中的含量为 0.7%~1.5%,蔬菜中南瓜的果胶含量最多,达到 7%~17%。其主要用途是用作酸性食品的胶凝剂、增稠剂等。果胶是一种分子中含有几百到几千个结构单元的线性多糖,平均相对分子质量大约在 50000~180000 之间,其基本结构是以 α-1,4-苷链结合的聚半乳糖醛酸,在聚半乳糖醛酸中,部分羧基被甲醇酯化,剩余的部分与钾、钠或铵等离子结合。高甲氧基化果胶分子的部分链节如下所示。

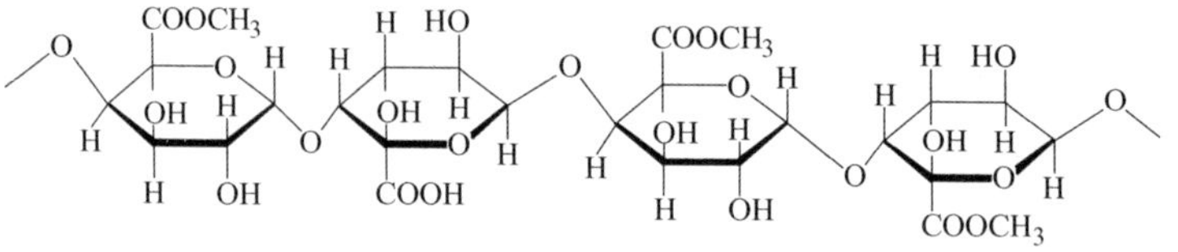

在果蔬中果胶多以原果胶存在。在原果胶中,聚半乳糖醛酸可被甲基部分酯化,并以金属桥(特别是钙离子)与多聚半乳糖醛酸分子残基上的游离羧基相连结,其结构为:

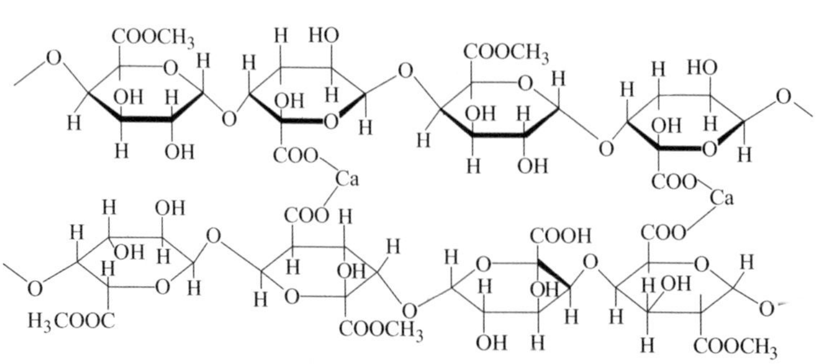

原果胶不溶于水,用酸水解时这种金属离子桥(离子键)被破坏,即得到可溶性果胶。再进行纯化和干燥即为商品果胶。

世界上柑橘年产量超过 $5×10^8$ t，其果皮约占 20%，为提取果胶提供了丰富的原料，也是目前我国常用的一种原料，所以本实验采用橘皮为原料，采用酸法萃取、酒精沉淀这一最简单的工艺路线来提取果胶。

【仪器与药品】

(1) 仪器：圆底烧瓶，漏斗，吸滤瓶。

(2) 药品：0.3%HCl 溶液，1%氨水，95%乙醇，柑橘皮，活性炭。

【步骤】

(1) 原材料预处理[1]：称取新鲜柑橘皮 5g 用清水漂洗干净，加水 20mL，加热到 90℃，保持 10min 以达到灭酶的目的。取出用水冲洗后切成小颗粒，再用 50~60℃的热水漂洗至漂洗水为无色，果皮无异味为止。

(2) 酸法萃取：将经过上述处理的果皮加入 0.3%HCl 溶液（刚好浸没果皮），其 pH 值控制在 2.0~2.5，加热到 90℃左右，提胶 50min，趁热用 100 目尼龙布（或四层纱布）过滤。

(3) 脱色[2]：在上述滤液中加入 2%的活性炭，在 80℃下加热 20min，以脱除色素和异味，趁热过滤[3]。

(4) 酒精沉淀：将上述滤液冷却，用 1%氨水调节 pH 值为 3~4，在不断搅拌下加入 95%的乙醇，使混合液中的乙醇浓度达 50%~60%，然后静置 10min，让果胶沉淀完全，用 100 目尼龙布过滤果胶，压干。将该果胶进一步用 95%乙醇处理，过滤，压干。

(5) 干燥：产品可在 105℃下烘干[4]。

【附注】

[1] 预处理的主要目的是灭酶，以防果胶发生酶解。同时也是对果皮进行清洗，以除去泥土、杂质、色素等。这种处理的好坏直接影响果胶的色泽和质量。

[2] 如果萃取液清澈透明，则可不用脱色。

[3] 因为胶状物容易堵塞滤纸，这时可加入占滤液量 2%~4%的硅藻土作助滤剂。

[4] 果胶凝胶试验：将 0.1g 柠檬酸和 0.08g 柠檬酸钠溶解于 10mL 水中，将果胶适量拌入 1~2 倍白糖中，然后加入到柠檬酸和柠檬酸钠的溶液中，不断搅拌，加热至沸，在白糖完全溶解后继续煮沸 20min，冷却后即成果酱。

Experiment 58 Extraction of Caffeine from Tea

【Experimental Objectives】

To extract caffeine from tea powder using polar-nonpolar solvent extraction technique.

【Experimental Principles】

Caffeine is a minor constituent of tea, coffee, and other natural plant materials.

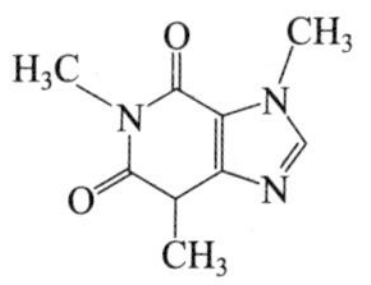

Caffeine

Extraction is a method used for the separation of organic compound from a mixture of compounds. This technique selectively dissolves one or more compounds into an appropriate solvent. The solution of these dissolved compounds is referred to as the extract. In the case of caffeine extraction from tea powder, the solubility of caffeine in water is 22mg/mL at 25℃, 180mg/mL at 80℃, and 670mg/mL at 100℃. Here the organic solvent dichloromethane is used to extract caffeine from aqueous extract of tea powder because caffeine is more soluble in dichloromethane (140mg/mL) than it is in water (22mg/mL). The dichloromethane-caffeine mixture can then be separated on the basis of the different densities of dichloromethane and water because dichloromethane is much denser than water and insoluble in it. Residual water is separated from dichloromethane by drain out the dichloromethane through separating funnel, thus dichloromethane passed through the funnel while polar solvents such as water is still remains in the funnel. Water and dichloromethane is slightly soluble in each other. So, after separating the solvents, residual water will remain the organic layer. Mainly anhydrous sodium sulfite is used for the removal of water from organic layer. Anhydrous sodium sulfite is an insoluble inorganic solid which will absorb water, thus drying it.

【Apparatus and Reagents】

(1) Apparatus: Beaker (500mL), hot plate, separating funnel.

(2) Reagents: Tea bags, dichloromethane, anhydrous sodium sulfite, distilled water.

【Experimental Procedure】

Take 1 tea bags and record the weight of these tea bags. Take 50mL beaker add 20mL of distilled water to it. Now place the tea bags in this beaker. Boil the contents in the beaker vigorously using a hot plate. Allow the mixture to cool for 5min and then decant the mixture into another beaker. Gently squeeze the tea bags to liberate the rest of the water. Cool the aqueous solution to near room temperature. Continue cooling in an ice box, the tea must be cool (20℃) before coming in contact with dichloromethane (boiling point=40℃). Extract the solution three times with 30mL portions of dichloromethane (CH_2Cl_2).

Extraction step: (a) The tea solution is poured into a separating funnel and 10mL of dichloromethane is added to it. The mixture will separate into two layers-the top layer is the tea layer and bottom layer is the dichloromethane since it is denser than tea. (b) Remove the funnel from the stand and keep your fingers on the stopper and carefully shake the separating funnel. (c) Vent the separating funnel periodically (every 30s) to relieve vapor pressure created inside the funnel. (d) When the contents have been sufficiently shaken place the separating funnel back on the ring stand and let the two layers separate. (e) Drain the bottom layer into a conical flask because now the caffeine is extracted into the dichloromethane layer. Cover the mouth of the conical flask to avoid evaporation of solution. (f) Repeat steps (a) through (e) twice.

Then, dry the combined dichloromethane solutions with anhydrous sodium sulfite. Add about 1 teaspoon of the drying agent until it no longer clumps together at the bottom of the flask. Mix well and leave it for 10min. Decant the dichloromethane into a conical flask (25mL). Evaporate the dichloromethane solvent in a hot water bath. When all the solvent is removed you observe a residue of yellowish green-white crystalline caffeine.

Sublimation step: (a) Take the conical flask containing crystalline caffeine. (b) Sublime the crude caffeine at atmospheric pressure by placing the flask directly on a pre-heated hot plate. Caffeine melts at 238℃ and sublimes at 178℃. (c) Collect your sublimed caffeine by keeping a test tube on the mouth of the conical flask. (d) White vapor of caffeine sticking onto the test tube and the walls of the conical flask is observed. (e) Now cool the conical flask.

Take a clean watch glass and record its weight in a weigh balance. Now strip off the caffeine from the conical flask and the walls of the test tube into the watch glass using a spatula. Record the weight of the watch glass + caffeine in a weigh balance and then find out the weight of extracted pure caffeine. The melting point of the extracted caffeine is determined using the melting point apparatus.

【Notes】
Do not get dichloromethane on your hands.

3.18　光学异构体的拆分

实验 59　苯乙胺的制备及外消旋体的拆分

（一）苯乙胺的制备

【目的与要求】
掌握 Leukart 反应的原理和外消旋体 α-苯乙胺的合成方法。

【原理】
反应原理：

$$HC(=O)-ONH_4 \rightleftharpoons HCOOH + NH_3$$

$$\text{>}C=O + NH_3 \xrightleftharpoons[]{-H_2O} \text{>}C=NH \xrightleftharpoons[]{NH_4^+} \text{>}C\overset{+}{-}NH_2$$

$$-O-\overset{O}{\underset{\|}{C}}-H + \text{>}C\overset{+}{=}NH_2 \longrightarrow CO_2 + H-\overset{|}{\underset{|}{C}}-NH_2$$

$$C_6H_5\overset{O}{\underset{\|}{C}}CH_3 + 2HCO_2NH_4 \longrightarrow C_6H_5\overset{CH_3}{\underset{|}{C}}H-NHCHO + NH_3\uparrow + CO_2\uparrow + 2H_2O$$

$$C_6H_5\overset{CH_3}{\underset{|}{C}}H-NHCHO + HCl + H_2O \longrightarrow C_6H_5\overset{CH_3}{\underset{|}{C}}HN\overset{+}{H_3}Cl^- + HCO_2H$$

$$C_6H_5\overset{CH_3}{\underset{|}{C}}HN\overset{+}{H_3}Cl^- + NaOH \longrightarrow C_6H_5\overset{CH_3}{\underset{|}{C}}HNH_2 + NaCl + H_2O$$

（±）-苯乙胺

【仪器与药品】

(1) 仪器：二口烧瓶，圆底烧瓶，微型分馏头，冷凝管，分液漏斗，吸滤瓶，温度计。

(2) 药品：苯乙酮，甲酸铵，苯，氢氧化钠，浓盐酸。

【步骤】

在二口烧瓶中加入 3.0g 苯乙酮、5g 甲酸铵和几粒沸石，在二口烧瓶的一侧装温度计，中间装上蒸馏头，装配成一个简单蒸馏装置。在烧瓶下加热，瓶内甲酸铵逐渐熔化分成两相，同时可见有液体慢慢蒸出。当反应混合物温度升至 150～155℃ 时，混合物变成均相。控制加热使反应混合物温度逐渐升高，到达 185℃ 时可停止加热[1]。馏出物中有少量苯乙酮，可用分液漏斗分出后倒回反应瓶中，重新加热反应，维持反应物温度不超过 185℃。

待反应混合物冷却后，倒入分液漏斗中，每次用 3mL 水洗涤 2 次。将有机层倒回二口烧瓶中，水层用苯萃取 2 次，每次 3mL。萃取液也倒入二口烧瓶中，加入 3mL 浓盐酸和几粒沸石，加热蒸馏除去苯，再改成回流装置，加热回流 0.5h。

反应混合物冷却至室温[2]，用苯萃取 2 次（洗去什么？），每次 10mL。将水层置于圆底烧瓶中，慢慢加入氢氧化钠溶液（2.5g 氢氧化钠＋4mL 水）进行水蒸气蒸馏。直至蒸出的馏液不呈碱性。馏出液用苯萃取 3 次，每次 4mL。苯萃取液用粒状氢氧化钠干燥后，进行简单分馏，收集 180～190℃ 的馏分，可得（±）-α-苯乙胺[3]，产量约为 1.6g。

纯（±）-α-苯乙胺沸点为 187.4℃。

【附注】

[1] 反应混合物温度不宜超过 185℃。在蒸馏过程中，蒸馏头与冷凝管中会析出固体碳酸铵，若析出固体太多会导致冷凝管堵塞，应停止加热，用水洗去碳酸铵后，重新加热。

[2] 若析出晶体，可加入少量水使之溶解。

[3] 游离的苯乙胺易吸收空气中的二氧化碳形成碳酸盐，故应塞好瓶口隔绝空气保存。

思 考 题

(1) 伯胺还可以用什么方法制备？

(2) 本实验中为什么在水蒸气蒸馏前要将溶液碱化？如不用水蒸气蒸馏，还可采取什么方法分离出游离的胺？

（二）外消旋体的拆分

【目的与要求】

(1) 掌握将外消旋体转变为非对映异构体后运用分部结晶的拆分方法。

(2) 掌握旋光仪的使用方法。

【原理】

合成的产物为外消旋体。要将外消旋的一对对映体分开，一般是将其与拆分剂形成非对映体，然后利用非对映体物理性质的不同，用结晶的方法将它们分离、精制，然后再去掉拆分剂，可得纯的旋光异构体。常用来形成非对映体的拆分剂有酒石酸、奎宁等。

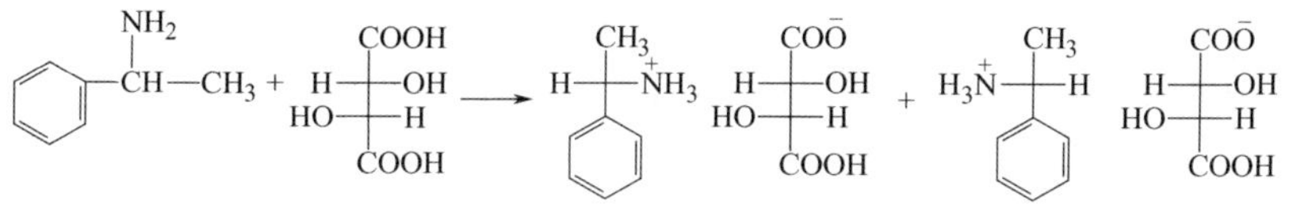

（±）-α-苯乙胺　　（+）-酒石酸　　（+）-α-苯乙胺-（+）-酒石酸盐　　（-）-α-苯乙胺-（+）-酒石酸盐

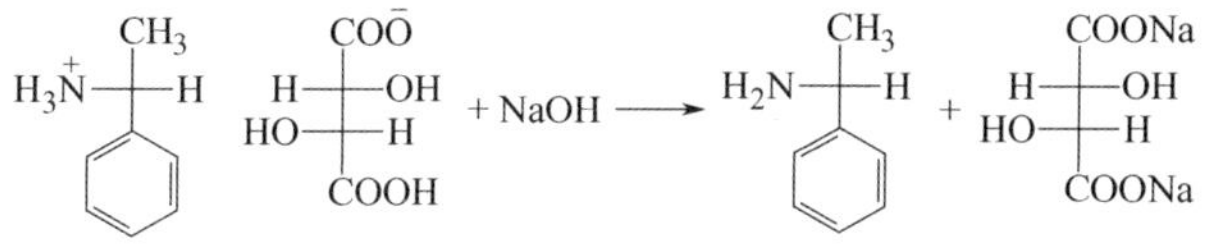

（-）-α-苯乙胺

【仪器与药品】

（1）仪器：锥形瓶，回流冷凝管，三角漏斗，吸滤瓶，圆底烧瓶，容量瓶，旋光仪。

（2）药品：（+）-酒石酸，甲醇，（±）-α-苯乙胺，乙醚，无水硫酸镁，氢氧化钠。

【步骤】

在锥形瓶中加入 1.5g（+）-酒石酸、25mL 甲醇和几粒沸石，装上回流冷凝管，在水浴上加热至沸腾，待（+）-酒石酸全部溶解后，稍冷，用滴管将 1.21g（±）-α-苯乙胺慢慢滴入热溶液中[1]。滴完后，稍加振摇，塞紧瓶塞，放置过夜。瓶内应生成颗粒状棱柱形晶体。若得到针状晶体与棱柱形晶体的混合物[2]，可将锥形瓶装上冷凝管后置于热水浴中加热，并不时振摇，因针状晶体较易溶解，少量棱柱形晶体留在溶液中作为晶种，再缓缓冷却溶液。待结晶完全后，抽滤，晶体用少量冷甲醇洗涤，空气干燥，所得产品主要是（-）-α-苯乙胺-（+）-酒石酸盐，1.0g。

将 1.0g 上述盐溶于 5mL 水中，加入 50% 氢氧化钠溶液 0.75mL，经充分振摇后，每次用 10mL 乙醚萃取 3 次。乙醚层用无水硫酸镁加 1 小粒氢氧化钠干燥 10min 左右。过滤，将滤液加到圆底烧瓶中，蒸除乙醚，减压下蒸馏得（-）-α-苯乙胺，其沸点为 84～85℃/3.5kPa，产量约为 0.2g。测定旋光度。纯（-）-α-苯乙胺的比旋光度 $[α]_D^{25}=-39.5°$。

【附注】

[1] 有时混合物会起泡和溢出，故操作时需小心。

[2] 针状晶体的存在使对映异构体拆分达不到足够高的光学纯度。将其转化成棱柱状结晶。

思　考　题

如何控制反应条件才能分离出纯的旋光异构体？

3.19　相转移催化和卡宾反应

实验 60　扁　桃　酸

【目的与要求】

（1）了解相转移催化的基本原理。

（2）掌握在相转移催化剂存在下，合成 d,l-扁桃酸的方法。

【原理】

在有机合成中常遇到有机相和水相之间的非均相反应。这种非均相反应的反应速度慢，产率低，甚至难以发生，同时副反应多，操作复杂。而相转移催化剂是使两种互不相溶的物质在一相中发生反应（如下所示）。因而反应速度快，反应条件温和，容易控制，产物便于分离。

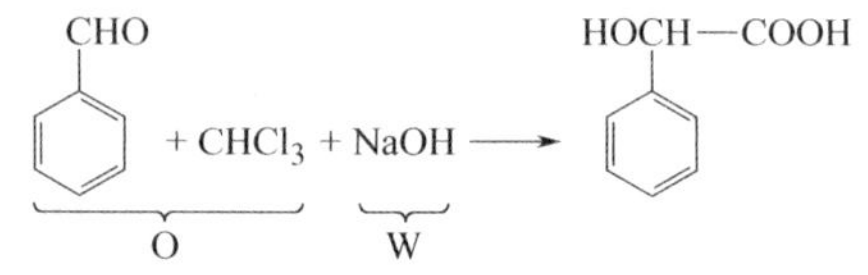

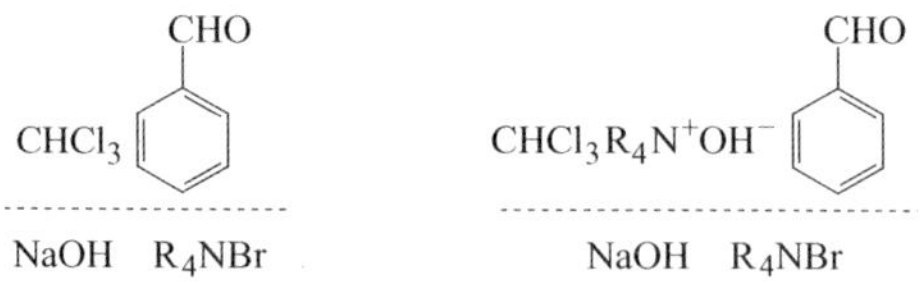

$$\text{PhCHO} + \text{CHCl}_3 + \text{NaOH} \longrightarrow \text{PhCH(OH)COOH}$$

d,l-扁桃酸（d,l-mandelic acid）化学名称为 α-羟基苯乙酸。作为药物可治疗尿路感染。在化学上可作为测定锆和铜的试剂，也是重要的有机合成中间体。

本实验的合成原理如下：

$$\text{CHCl}_3 + \text{R}_4\text{N}^+\text{OH}^- \rightleftharpoons \text{R}_4\text{N}^+\overline{\text{CCl}}_3 + \text{H}_2\text{O}$$

$$\text{R}_4\text{N}^+\overline{\text{CCl}}_3 \rightleftharpoons \text{R}_4\text{N}^+\text{Cl}^- + :\text{CCl}_2$$

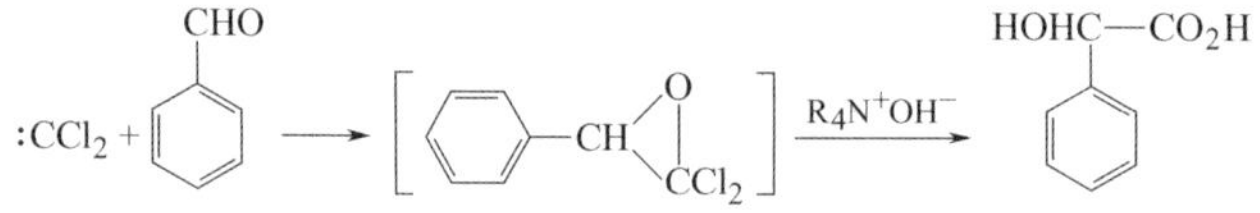

$$:\text{CCl}_2 + \text{PhCHO} \longrightarrow \left[\text{Ph-CH}\underset{\text{CCl}_2}{\overset{\text{O}}{\diagdown}}\right] \xrightarrow{\text{R}_4\text{N}^+\text{OH}^-} \text{PhCH(OH)CO}_2\text{H}$$

【仪器与药品】

(1) 仪器：25mL 三口烧瓶，分液漏斗，滴液漏斗。

(2) 药品：苯甲醛，乙醚，氢氧化钠，氯化三乙基苄铵（TEBA），甲苯，氯仿（50%硫酸，无水硫酸钠）。

【步骤】

在三口烧瓶中加入 5mL 苯甲醛，0.5g TEBA、8mL 氯仿，搅拌均匀，升温至 56℃ 滴加氢氧化钠溶液，温度维持在 60～65℃（不得超过 70℃，滴加约需 0.5h），加完后继续反应 1h，温度控制在 65～70℃，当反应液 pH 值至近中性时，停止反应。将反应液用 100mL 水吸收稀释，用乙醚萃取 2 次，每次 10mL，合并醚层，待回收用[1]。水层用 50%硫酸酸化至 pH 值为 2～3，用乙醚萃取 3 次，每次 20mL，合并提取液，并用无水硫酸钠干燥，蒸馏除去乙醚，得粗产物，其产量约为 5.5g。粗产物可用甲苯重结晶。产品呈白色结晶，m.p.：118～119℃。

【附注】

[1] 这里首先萃取的是原料和副产物。

实验 61　1-苯基-2,2-二氯环丙烷

【目的与要求】

(1) 了解相转移催化的基本原理。

(2) 掌握在相转移催化剂存在下，合成 1-苯基-2,2-二氯环丙烷的方法。

【原理】

氯仿与氢氧化钠水溶液在相转移催化剂的存在下反应生成二氯卡宾，与苯乙烯发生加成反应，生成 1-苯基-2,2-二氯环丙烷。

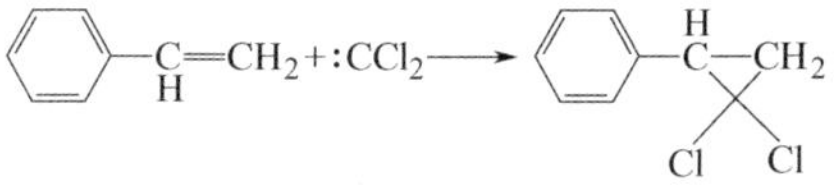

$$\text{PhCH=CH}_2 + :\text{CCl}_2 \longrightarrow \text{PhCH-CH}_2\text{-CCl}_2$$

【仪器与药品】

（1）仪器：25mL 三口烧瓶，分液漏斗，冷凝管。

（2）药品：苯乙烯，氯化三乙基苄铵，氢氧化钠，氯仿。

【步骤】

按图 1-5 装置仪器，在 25mL 的三口烧瓶中加入新鲜蒸馏的苯乙烯 5.2g、氯仿 6g、50%氢氧化钠水溶液 10mL 及氯化三乙基苄铵 0.2g，于 40℃下搅拌 4h。将反应液倒入水中，分出有机层，有机层经水洗、干燥和过滤，蒸去低沸物，最后减压蒸馏，收集 78~83℃/266.6Pa 的馏分。

Experiment 62　Synthesis of 7,7-Dichlorobicyclo [4.1.0] heptane

【Experimental Objectives】

（1）Understand the principle of phase transfer catalytic reaction.

（2）Learn and master the operations of mechanical stir, extraction, vacuum distillation and Abbe refractometer.

【Experimental Principles】

Dichlorocarbene is the reactive intermediate with chemical formula CCl_2. Although this material has not been isolated, it is a common intermediate in organic chemistry, being generated from chloroform.

Dichlorocarbene is most commonly generated by reaction of chloroform and a base such as potassium t-butoxide or aqueous sodium hydroxide. A phase transfer catalyst, for instance benzyltriethylammonium bromide, facilitates the migration of the hydroxide in the organic phase.

Addition of carbene to cyclohexene under PTC conditions afford 7,7-dichlorobicyclo [4.1.0] heptanes.

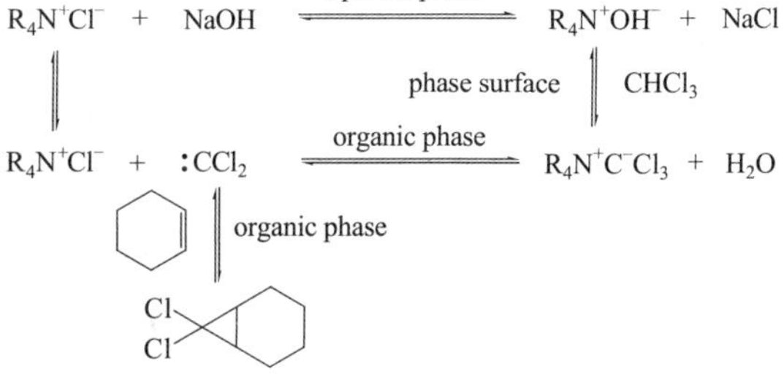

$$R_4N^+Cl^- + NaOH \xrightleftharpoons{\text{aqueous phase}} R_4N^+OH^- + NaCl$$

$$\Big\updownarrow \qquad \text{phase surface} \quad \Big\updownarrow CHCl_3$$

$$R_4N^+Cl^- + :CCl_2 \xrightleftharpoons{\text{organic phase}} R_4N^+C^-Cl_3 + H_2O$$

【Apparatus and Reagents】

（1）Apparatus: Cyclohexene, chloroform, triethylbenzylammonium chloride（TEBA, phase transfer catalyst), NaOH, hydrochloric acid, saturated NaCl solution, diethyl ether.

(2) Reagents: Three-necked round bottom flask, spherical condenser, thermometer, separatory funnel.

【Experimental Procedure】

Cyclohexene (2.02mL), TEBA (0.1g) and chloroform (6mL) are added to a 25mL three-necked round bottom flask. Stir and add 50% NaOH solution (6mL) slowly from the top of the condenser within 15min.

After the addition of NaOH solution, continue stirring the mixture for 1.5h without heating. Control the temperature at 50—60℃. The reaction is stopped automatically, that is, the temperature of the reaction system drops gradually to the room temperature. After adding 15mL water, the mixture is diluted and transferred into a separatory funnel, and the organic phase is separated. The water phase is extracted with ether (5mL×3). Combine the organic phase and ether, wash the organic solution with hydrochloric acid solution (2mol/L, 10mL) and saturated NaCl solution (5mL×3), then dry liquid over magnesium sulfate for 0.5—1h.

Filtrate and transfer the liquid into a round bottom flask, remove the ether, chloroform and unreacted cyclohexene by atmospheric distillation. Then obtain the final product by vacuum distillation. Collect the boiling fractions at 89—91℃/2.27kPa, 95—97℃/4.65kPa. The yield is 1.6—2g.

3.20 碳水化合物及其衍生物

实验63　五乙酸葡萄糖酯

【目的与要求】

(1) 了解葡萄糖的开链和半缩醛的性质。

(2) 掌握五乙酸葡萄糖酯的合成方法。

【原理】

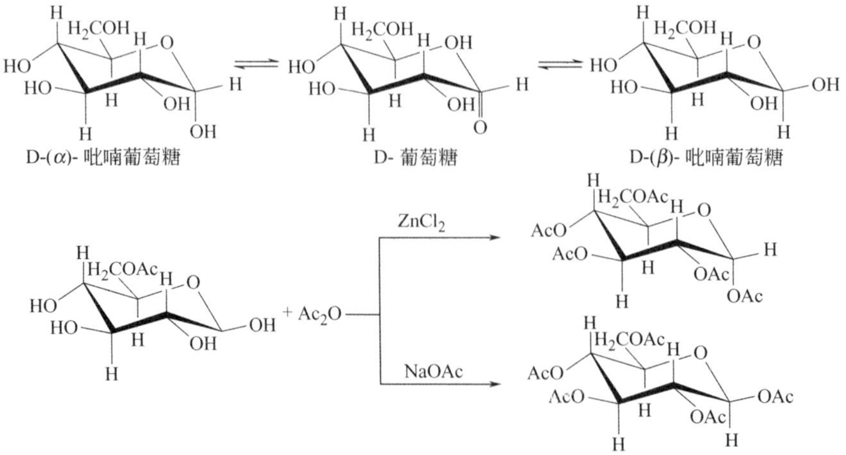

α-五乙酸葡萄糖酯的熔点为110～111℃，比旋光度 $[\alpha]_D^{25} = -101°$；β-五乙酸葡萄糖酯

的熔点为 131~132℃，比旋光度 $[\alpha]_D^{25}=-4°$。

【仪器与药品】

(1) 仪器：25mL 三口烧瓶，漏斗，温度计，H 形蒸馏头。

(2) 药品：无水氯化锌，乙酸酐，无水乙酸钠，葡萄糖，甲醇，无水乙醇。

【步骤】

(1) 五乙酸-D-α-葡萄糖酯的制备：向装有温度计、H 形蒸馏头的 25mL 三口烧瓶中加入 6mL 乙酸酐和研细的无水氯化锌[1]，微微加热并搅拌至全溶。冷却后在搅拌下缓慢加入 1g 葡萄糖[2]，再逐渐升温到 80~85℃，搅拌反应 1h。稍冷后，将反应物在搅拌下倾入 100g 冰水中，立即有黄棕色黏稠物析出，继续搅拌至固化，过滤，水洗，干燥，甲醇重结晶，得无色晶体。

(2) 五乙酸-D-β-葡萄糖酯的制备：装有温度计、H 形蒸馏头的 25mL 三口烧瓶中加入 1.5g 无水乙酸钠[3]和 12.3mL 乙酸酐，水浴加热，搅拌 10min，稍冷后在搅拌下缓慢加入 2.5g 葡萄糖，于沸水中加热搅拌 2.5h。稍冷后将反应物在搅拌下倾倒入 200g 冰水中，立即有黄棕色黏稠物析出，继续搅拌至固化，过滤，水洗，干燥，用 20mL 乙醇重结晶，得无色晶体。

【附注】

[1] 氯化锌应经过加热至熔融，放在干燥器中冷却固化，研细，贮存于干燥器中备用。

[2] 所用的葡萄糖需经过真空干燥箱的充分干燥。

[3] 无水乙酸钠应该自行预处理：将含结晶水的乙酸钠放入坩埚中，加热融化，并蒸发除去结晶水，逐渐成为固体，进一步加热使其重新融化，再小心加热几分钟。停止加热后立即将其研细，贮存于磨口瓶中备用。

实验 64　从牛奶中制取酪蛋白和乳糖

【目的与要求】

掌握从牛奶中制取酪蛋白和乳糖的原理和方法。

【原理】

牛乳中含有 3.0%~3.7% 的蛋白质，其中约有 80% 为酪蛋白。酪蛋白以酪蛋白酸钙、磷酸钙的络合物胶粒状态在牛奶中形成胶体溶液。酪蛋白是一种含磷蛋白质，其元素组成为 C，53.07%；H，7.13%；N，15.64%；P，0.8%；S，0.76%，在酪蛋白形式的胶束中还会结合柠檬酸、镁等物质，具有比较复杂的结构。和其他蛋白质一样，当到达等电点时，酪蛋白的胶体微粒的一系列物理性质都会发生很大的变化。由于这时溶胶的稳定性最小，酪蛋白很容易发生聚集而沉淀，从乳中分离酪蛋白正是利用了这一特性。酪蛋白的等电点为 pH＝4.6~4.7。用酶或酸都可以使酪蛋白凝聚。用酶时，钙离子不易除去，产品灰分高。用酸发生如下反应，容易将酪蛋白上的盐类除掉，减少干酪素中的灰分。

酪蛋白酸钙·$Ca_3(PO_4)_2$ + 4HCl ⟶ 酪蛋白 + $2Ca(H_2PO_4)_2$ + $2CaCl_2$

制得的食品干酪素可广泛应用于面包和灌肠制品中。此外若将酪蛋白中的乳糖和盐类除去，并使酪蛋白转变为酪蛋白酸钠和酪蛋白酸钙后，可用作食品乳化剂。

牛乳分离掉酪蛋白后，其中乳糖占其干物质的 65%~74%，乳清经真空浓缩 10~12 倍便可结晶出粗乳糖，再经过脱色、重结晶等可制得乳糖。乳糖常用于婴儿食品、糖果、人造奶油等。此外在医药上还用作矫味剂。

【仪器与药品】

(1) 仪器：离心分离器，10mL 烧杯、30mL 烧杯，蒸发皿。

(2) 药品：牛奶（或脱脂奶粉），乙酸，乙醇，乙醚，氢氧化钠，硝酸，碳酸镁。

【步骤】

将牛奶（或脱脂奶粉）放在离心分离器中，以 2000～3000r/min 的转速离心分离 15～20min，除去脂肪，脱脂牛奶供下面的实验使用。或取 1.5g 脱脂奶粉加入 40℃ 的水 40mL，搅拌溶解，备用。

在搅拌下，在上述溶液中慢慢加入预热到 40℃ 的 0.1% 的乙酸，调节 pH 值到 4.6～4.7，静置，离心，沉淀为酪蛋白粗品，清液供提取乳糖用。将该粗品加水充分搅拌，以洗涤除去其中的水溶性杂质（乳清蛋白、乳糖等），离心后上层清液并入前述清液。重复上述洗涤 1 次。所得的沉淀分别用 5mL 乙醇、乙醚洗涤，离心除去上层清液，以除去磷脂类物质和脂肪类物质。将沉淀摊开晾干，得酪蛋白纯品。

将上述清液煮沸，使水溶性的蛋白质变性沉淀，然后过滤除去，在滤液中加入碳酸镁中和，放在沸水浴上蒸出全部水分。向该干物质中加入乙醇，搅拌提取醇溶性物质后，尽量倾出上层溶液，在醇不溶物中加水（1g 加水 5mL）使其溶解过滤后，再在沸水浴上浓缩成糖浆状，放置让乳糖结晶析出，过滤，洗涤。

实验 65　食品增稠剂、保型剂乙酯淀粉的制备

【目的与要求】

(1) 了解淀粉改性的有关知识和常用方法。

(2) 掌握乙酯淀粉的有关实验方法。

【原理】

淀粉是天然高分子材料，具有增稠、胶凝、黏合和成膜等特性，并且还具有价廉、易得、质量容易控制等优点，因而广泛用于黏合剂、纺织、造纸、食品、医药以及建筑材料等产品的制造中。可用化学、物理和酶处理等方法来改进淀粉的特性，以强化或抑制它的原有特性或增添新的性质。

乙酯淀粉属于改性淀粉之一（其他还有羧甲基淀粉、氧化淀粉、磷脂淀粉、甲基淀粉等），具有较低的凝胶温度、溶液较稳定等优点，而有利于用作食品增稠剂、保型剂，还可用于纸面施胶、纺织品精加工等。

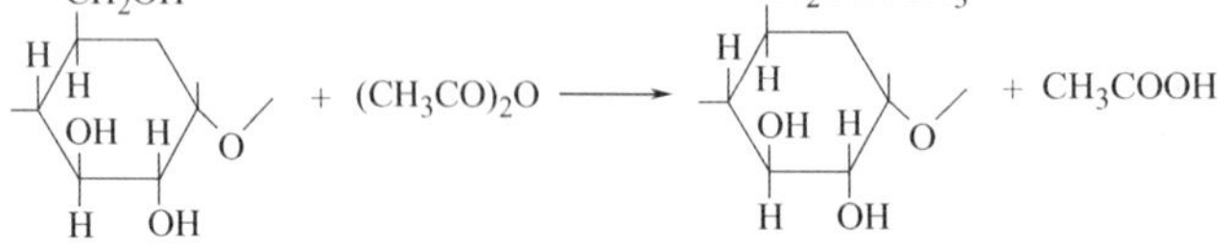

$$\text{淀粉-OH} + (CH_3CO)_2O \longrightarrow \text{淀粉-OOCCH}_3 + CH_3COOH$$

【仪器与药品】

(1) 仪器：25mL 三口烧瓶，电动搅拌器。

(2) 药品：玉米淀粉，乙酸酐，3% 氢氧化钠，稀盐酸。

【步骤】

在反应器中加入玉米淀粉 6g、水 7mL，搅拌均匀，在 25℃ 下加入 3% 氢氧化钠溶液，

调节 pH 值至 8 左右，然后慢慢滴加乙酸酐，同时滴加氢氧化钠溶液，保持 pH＝8.0～8.4。滴加完毕，继续搅拌 0.5h，用稀盐酸调节 pH＝4.5，过滤，用少量的水洗涤，干燥。

3.21 氨基酸与肽

实验 66　L-酪氨酸甲酯（L-Tyrosine methyl ester）

【目的与要求】

掌握 L-酪氨酸甲酯的合成方法。

【原理】

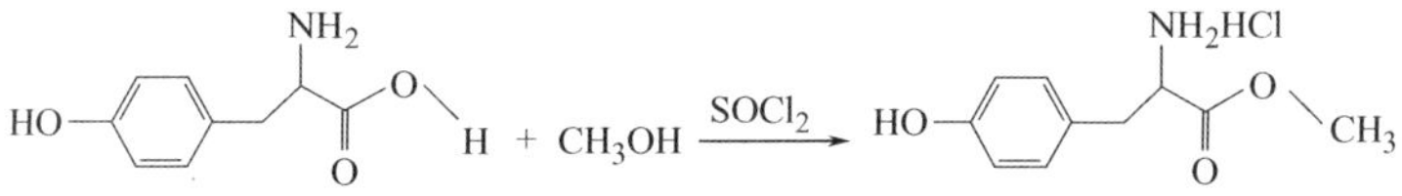

$$\text{HO-C}_6\text{H}_4\text{-CH}_2\text{-CH(NH}_2\text{)-COOH} + \text{CH}_3\text{OH} \xrightarrow{\text{SOCl}_2} \text{HO-C}_6\text{H}_4\text{-CH}_2\text{-CH(NH}_2\cdot\text{HCl)-COOCH}_3$$

【仪器与药品】

(1) 仪器：25mL 圆底烧瓶，滴液漏斗，砂芯漏斗。
(2) 药品：L-酪氨酸，氯化亚砜，无水甲醇，无水乙醚。

【步骤】

在 25mL 圆底烧瓶中加入 8mL 无水甲醇，在冰盐浴上冷却到 －10℃，在此温度下慢慢滴加 1mL 氯化亚砜，然后加入 2g L-酪氨酸，在 20℃下放置 2h，再在 70～80℃回流 30min，使全部 L-酪氨酸溶解。改用减压蒸馏装置，将部分甲醇蒸去。滴入 30～40mL 无水乙醚，结晶析出，并在冰盐浴上冷却 1h，待结晶完全，抽滤，再用无水乙醚洗涤结晶。可得 2.3g 产品，其熔点为 190℃，产率为 90%，比旋光度 $[\alpha]_D^{23}=+74℃$（吡啶）。

3.22 非常规有机合成方法

实验 67　化学发光指示剂 3-氨基邻苯二甲酰肼

【目的与要求】

(1) 了解化学发光的原理和用途。
(2) 掌握化学发光指示剂 3-氨基邻苯二甲酰肼的合成方法。

【原理】

一般而言，物体在受到高能辐射、经受高压放电或激烈的化学反应（燃烧）时才有发光现象产生。所谓化学发光（chemiluminescence），是指在化学反应过程中，无须加热，也无明显放热现象而产生可见光的一类反应。该反应中通常有一个产物的分子处于电子激发态，这种激发态分子随即发射光子，产生可见光。化学发光剂主要用于易燃、易爆，不能使用电、热光源的场合，如矿井、油田和军事等。此外在容量分析时，若在滴定溶液中加入了在滴定终点时能发光的物质，则达到滴定终点时，该指示剂就可使全部溶液出现可见的光亮，无论滴定溶液的背景是浑浊的还是深色的，都能正确而明亮地指示滴定终点的到来。这种发光剂叫做化学发光指示剂。

化学发光指示剂主要有肼类，多聚硅烃的氧衍生物等。3-氨基邻苯二甲酰肼（露明诺，Luminol）属于肼类，是应用较广的一种。一般认为其化学发光的过程如下：

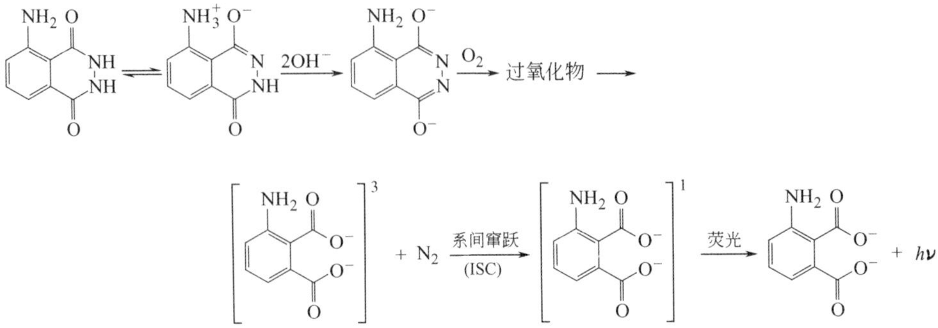

三线态双阴离子（T_1）　　单线态双阴离子（S_1）　　基态双阴离子

上述过程中的过氧化物的结构有多种假设，其中为多数人认可的是：

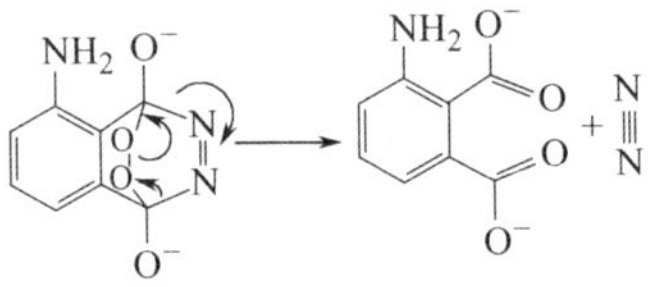

合成方法：

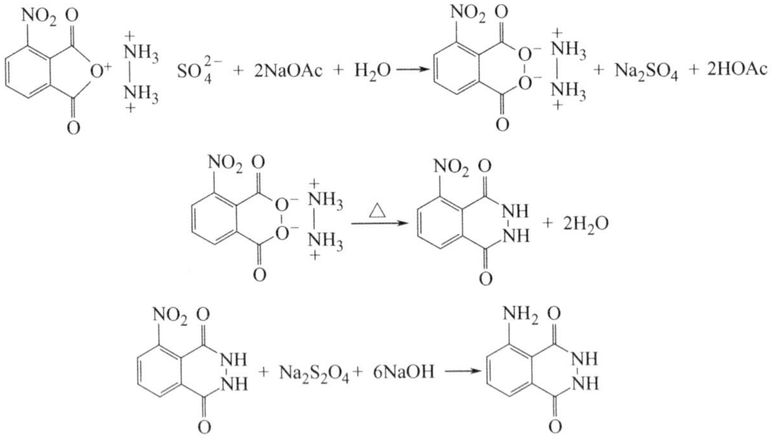

【仪器与药品】

（1）仪器：25mL 试管，砂芯漏斗。

（2）药品：3-硝基邻苯二甲酸酐，硫酸肼，结晶乙酸钠，保险粉 $Na_2S_2O_4$，氢氧化钠，冰醋酸，甘油，赤血盐，3％H_2O_2。

【步骤】

（1）酰肼的合成　3-硝基邻苯二甲酸酐 1g、硫酸肼 0.7g、结晶乙酸钠 1g 和水 4mL 依次加入 25mL 试管中，缓慢加热煮沸，让 HOAc 和水蒸气逸出，维持 15min。冷却片刻，加入 5mL 甘油和沸石，插入温度计，继续加热使水分逸出，并不时搅拌，防止崩沸，当温度升高到200～220℃时，在此温度下保持 3～4min，此时反应液呈黄色。试管冷却到 100℃以下，加入水 40mL，加热至沸腾，冷却，酰肼凝集，用吸管小心将水吸出。再加入 40mL 水，重复进行上述操作，得到酰肼。

(2) 还原　向上述酰肼的试管中加入水少许及 5mL 5% 氢氧化钠溶液，得橘红色溶液，分批加入保险粉 $Na_2S_2O_4$ 3g，加热至沸，维持 3～4min，溶液变成黄色，并有黄色产品 3-氨基邻苯二甲酰肼析出。冷却后加入冰醋酸酸化，使产品沉淀完全。抽滤，水洗，干燥。

（3）化学发光试验　试液 A：0.2～0.3g 产品溶解于 10mL 氢氧化钠溶液中，再加 90mL 水稀释。试液 B：20mL 赤血盐和 20mL 3% H_2O_2 混合，再加入 160mL 水稀释。

在暗室中将试液 A 和试液 B 以相同的速度加入锥形瓶中，并振摇，混合液即发出醒目的淡蓝色的光线。

【备注】

本实验可与其他某个实验同时做，在第一次实验时仅需 15min，并在随后约 1 星期后的下次实验课里再花 15min 即可。

实验 68　二苯甲酮的光化学还原

【目的与要求】

（1）了解光化学还原的原理和用途。
（2）掌握二苯甲酮的光化学还原制备苯频哪醇的方法。

【原理】

二苯甲酮的光化学还原是最老和研究得最彻底的光化学反应之一，反应式如下：

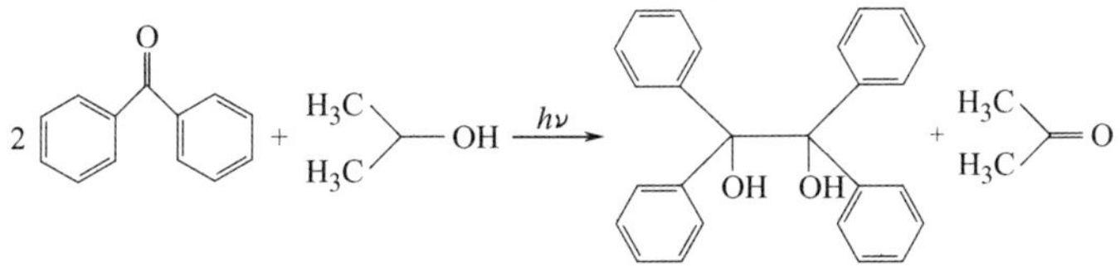

二苯甲酮的光化学还原是二苯甲酮的 n→π* 三线态（T_1）的反应，具体如下：

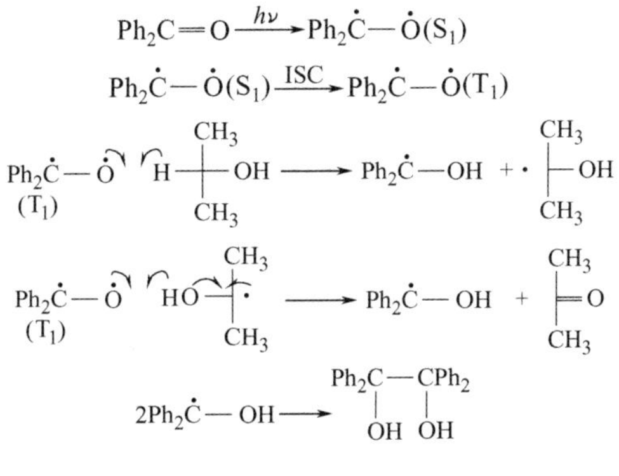

$$Ph_2C=O \xrightarrow{h\nu} Ph_2\dot{C}-\dot{O}(S_1)$$

$$Ph_2\dot{C}-\dot{O}(S_1) \xrightarrow{ISC} Ph_2\dot{C}-\dot{O}(T_1)$$

$$2Ph_2\dot{C}-OH \longrightarrow \underset{OH\ OH}{Ph_2C-CPh_2}$$

【仪器与药品】

（1）仪器：25mL 试管，砂芯漏斗。
（2）药品：二苯甲酮，异丙醇，冰醋酸。

【步骤】

将 2.5g 二苯甲酮和 10mL 异丙醇加入试管中，温热溶解，当固体溶解后，加入 1 滴冰醋酸，然后加入异丙醇至管顶，用橡皮塞将试管塞紧，并充分振摇，将试管置于窗台上的烧杯中，接受阳光的直射。经过 1 周左右，产物从溶液中析出，过滤收集产物，干燥。m.p.：188～190℃。

思 考 题

当二苯甲酮用金属镁处理时，发生一个与本实验类似的反应：

$$2\,Ph_2C{=}O \xrightarrow{Mg} \underset{\underset{OH}{|}}{Ph_2C}-\underset{\underset{OH}{|}}{CPh_2}$$

该反应的机理与光化学还原机理有何差别？

实验 69　微波辐射下的 Perkin 反应——肉桂酸的制备

【目的与要求】

(1) 熟悉微波反应的基本反应装置。

(2) 掌握微波条件下苯甲酸酯的制备方法。

【原理】

微波是频率大约在 300MHz～300GHz，即波长在 100cm～1mm 范围内的电磁波。其波长位于电磁波谱的红外和无线电波之间。微波作为化学反应中的热源具有三大特点：快速加热（即使存在大量离子）；分子水平意义上的搅拌，耗时短，效率高；升温速度可以在很大范围内变动。由于微波可使常规反应的速度千百倍地加快，并能明显提高反应得率，所以自 1986 年 Gedye 首次将商用微波炉用于苯甲酸的酯化反应以来（R. Gedye. et. al. Tetrhdron Lett. 1986，27，279.），微波技术被广泛用于 Diels-Alder 反应（R. J. Giguere, et. al. Tetrahedron Lett. 1986，27，4945.）、酯化、O-烃基化、Adol 反应、Reformastky 反应、重排反应、加成、催化氢化以及多种形成杂环的反应，可以说微波几乎涉及了有机合成反应的各个主要领域。但微波化学研究仍停留在实验事实的积累方面，对微波加速或改善化学反应的机理还没有给出令人信服的解释。

反应式：

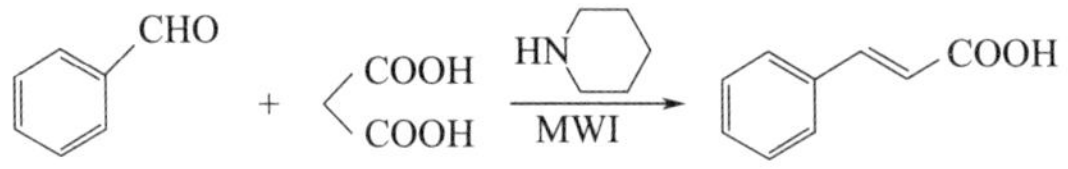

$$\text{PhCHO} + \text{CH}_2(\text{COOH})_2 \xrightarrow[\text{MWI}]{\text{HN}\langle\text{piperidine}\rangle} \text{PhCH=CHCOOH}$$

【仪器与药品】

(1) 仪器：微波反应器，三口烧瓶，球形冷凝管，布氏漏斗，抽滤瓶。

(2) 药品：苯甲醛，丙二酸，哌啶。

【步骤】

苯甲醛 1mmol，丙二酸 1.5mmol，哌啶 1.5mmol，混合置于 660W 微波下辐射 15s，化合物倾入冰中，稀盐酸酸化至 pH=4，过滤，水洗，用乙醇重结晶，给出肉桂酸，收率＞90%。m.p.：131～132℃。

思 考 题

检索文献，了解非微波条件下苯甲醛与丙二酸缩合的反应条件，并比较与体会其异同。

实验 70　微波辐射下苯并吡喃-2-酮-3-甲酸乙酯的制备

【目的与要求】

(1) 熟悉微波反应的基本原理及微波反应的操作方法。

(2) 掌握利用 Knoevenagel 缩合反应由水杨醛制备香豆素类化合物的方法。

【原理】

香豆素类化合物是一类重要的有机化合物，常被用作食品和饲料的添加剂、荧光增白剂、激光染料等。通过水杨醛与丙二酸酯类化合物的 Knoevenagel 缩合反应是获得此类化合物的一种重要方法。

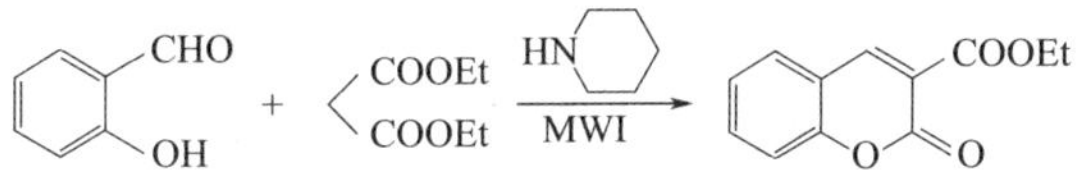

$$\text{邻羟基苯甲醛} + \begin{matrix}\text{COOEt}\\\text{COOEt}\end{matrix} \xrightarrow[\text{MWI}]{\text{HN}\bigcirc} \text{香豆素-COOEt}$$

【仪器与药品】

(1) 仪器：微波反应器，三口烧瓶，球形冷凝管，布氏漏斗，抽滤瓶。
(2) 药品：水杨醛，丙二酸二乙酯，哌啶，乙醇。

【步骤】

水杨醛 100mmol，丙二酸二乙酯 110mmol，哌啶 0.2g（2.4mmol），置于功率为 60～100W 的微波反应器中辐射 10min，最终体系温度达到 129℃，粗品直接以乙醇重结晶，得到产品，收率：80%～90%，m.p.：91～93℃。

思 考 题

请检索文献，了解苯并吡喃-2-酮-3-甲酸乙酯的应用。

实验 71 超声条件下 3-烯丙基-2,4-戊二酮的制备

【目的与要求】

(1) 了解超声化学的基本原理。
(2) 了解超声反应的基本反应装置。
(3) 掌握 3-烯丙基-2,4-戊二酮的制备方法。

【原理】

广义地说，声波包括次声波（＜20Hz）、声波（20Hz～20kHz）、超声波（20kHz～50MHz）以及微波超声（＞50MHz）。超声波的频率要比声波的高。与普通声波相比，超声波具有更好的束散性；超声波在介质中具有更强的吸收，其传播距离往往很短，超声波的频率越高，吸收越强，传播距离越短。一般而言，超声波在气相中吸收最强，传播距离最短，固相中吸收最弱，传播距离最长。如频率为 1000kHz 的超声波在空气中经过 0.5m 的距离其强度就衰减为原来的一半，在水中传播时经过 500m 的距离才衰减一半。

超声合成于 20 世纪 80 年代初取得进展，许多有机合成反应在施加超声辐射后，可以在几十乃至几分钟内迅速完成，且副反应减少，是有机合成中非常有实用价值的实验手段。人们把在超声辐射下的化学反应及有关工作统称为声化学（Sonochemistry）。

超声能够促进常规反应的原因主要是：超声波具有比声波大得多的能量，可以供给介质分子足够大的能量，这种能量可以使介质的质点获得极大的加速度（可以达到重力加速度的几十万倍乃至几百万倍）；超声波可以使介质质点产生显著的声压，这种附加压力可达几巴（1bar=10^5Pa）；超声波在作用于液体介质时还产生一种异常重要的效应——空化作用。空化气泡内的瞬间温度可达到数千摄氏度，同时存在巨大的压力和冲击波。

超声波的频率接近于一般金属晶格振动的频率，因而对金属参加的固-液相反应特别

有效。

对有机反应施加超声辐射通常有两种方法：一种简易的方法是将反应器置于超声清洗器中，但这种方法产生的超声辐射必须经过槽中的介质传递，损耗大；且介质用量、反应器大小及处于槽中的位置等均可对反应结果产生影响，所以重现性也较差。另一种方法是将超声探头直接插入反应器中，可避免第一种方法的缺点。

超声波可以促进常规反应，提高反应产率和选择性，缩短反应时间，甚至可以激发通常情况下不能发生的化学反应。在工业上有可能实现管道化连续生产，因而具有较为广阔的应用前景。

合成方法：

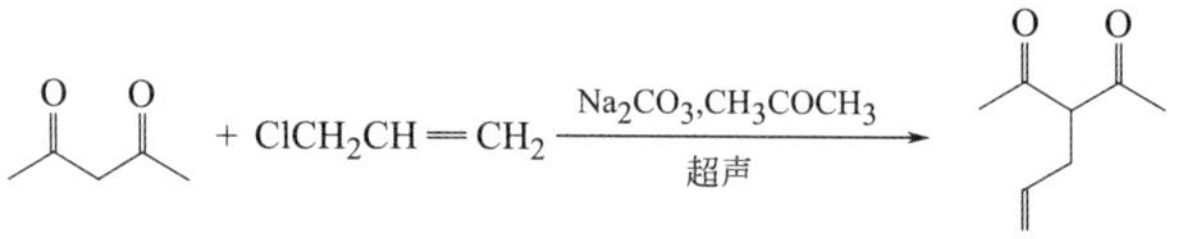

$$\underset{}{\text{乙酰丙酮}} + ClCH_2CH=CH_2 \xrightarrow[\text{超声}]{Na_2CO_3, CH_3COCH_3} \text{3-烯丙基-2,4-戊二酮}$$

【仪器与药品】

（1）仪器：锥形瓶，超声反应器，布氏漏斗，抽滤瓶，水循环真空泵。

（2）药品：乙酰丙酮，无水碳酸钠，丙酮，氯丙烯。

【步骤】

于500mL锥形瓶中加入2.0g（20mmol）乙酰丙酮、2.6g（24.5mmol）无水碳酸钠和50mL丙酮氯丙烯22mmol，将反应瓶置于超声波发生器中，开动超声波在室温下反应1.5h，过滤固体并用丙酮洗涤三次，合并滤液和洗涤液，水泵减压蒸去溶剂，得到粗产物，产率70%～85%。

思 考 题

请检索文献，了解3-烯丙基-2,4-戊二酮的合成方法，比较常规方法与微波、超声条件下费用的不同，并叙述本实验方法的先进性。

Experiment 72 Synthesis of Methyl Benzoate under Ultrasonic Radiation

【Experimental Objectives】

Understand the ultrasonic radiation can improve the reaction conditions and short the reaction time

【Experimental Principles】

Under the ultrasonic radiation condition, benzoic acid can react with methanol to form ester compound with the p-toluenesulfonic acid as a catalyst. This reaction condition is mild. The product selectivity and yield could be enhanced compared with the traditional high temperature and high pressure reaction conditions.

$$C_6H_5COOH + CH_3OH \xrightarrow[\text{ultrasonic}]{H_3C-C_6H_4-SO_3H} C_6H_5COOCH_3$$

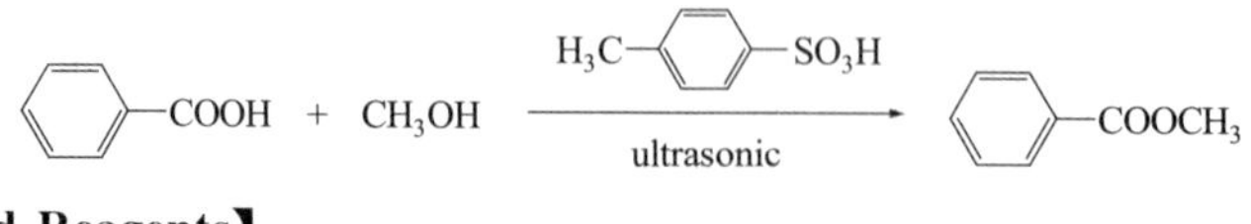

【Apparatus and Reagents】

(1) Apparatus: 350W ultrasonic instrument, flask, condenser pipe.

(2) Reagents: Benzoic acid, methanol, p-toluenesulfonic acid, 10% NaHCO$_3$ solu-

tion.

【Experimental Procedure】

In 25mL round-bottom flask, 0.05mol benzoic acid and 0.35mol methanol are added. *p*-toluenesulfonic acid, as a catalysis, is added 8% quality dosage of the benzoic acid. Then, the flask is put into the ultrasonic instrument and heated to 80℃. The ultrasonic power is set as 350W to treat the mixture for 70min. After cooling to the room temperature, the solution is adjusted the pH to 7—8 by adding 10% $NaHCO_3$ solution. The organic layer is collected and dried. The solvent is removed by the reduced pressure distillation to afford the colorless transparent oily liquid.

【Notes】

The b.p. of methyl benzoate is 198℃.

Experiment 73 Synthesis of 1,2,3,4-Tetrahydrocarbazole

【Experimental Objectives】

(1) Understand the microwave heating can improve the reaction conditions and short the reaction time.

(2) The mechanism of synthesis of 1,2,3,4-tetrahydrocarbazole.

【Experimental Principles】

Phenylhydrazine condensation with cyclohexanone catalysis by acetic acid afford hydrazone, then the hydrazone takeplace [3,3]-sigmatropic rearrangement give the product 1,2,3,4-tetrahydrocarbazole.

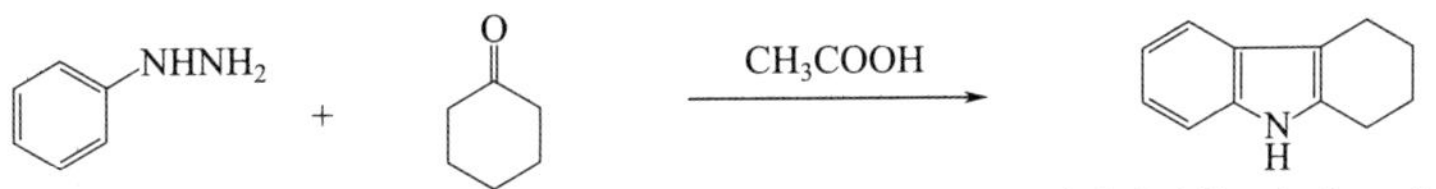

	phenylhydrazine	cyclohexanone		1,2,3,4-Tetrahydrocarbazole
MW:	108.14	98.14		171.24
mass/g	1.08	1.47		1.712(theory)
moles:	0.01	0.015		0.01(theory)

Mechanism:

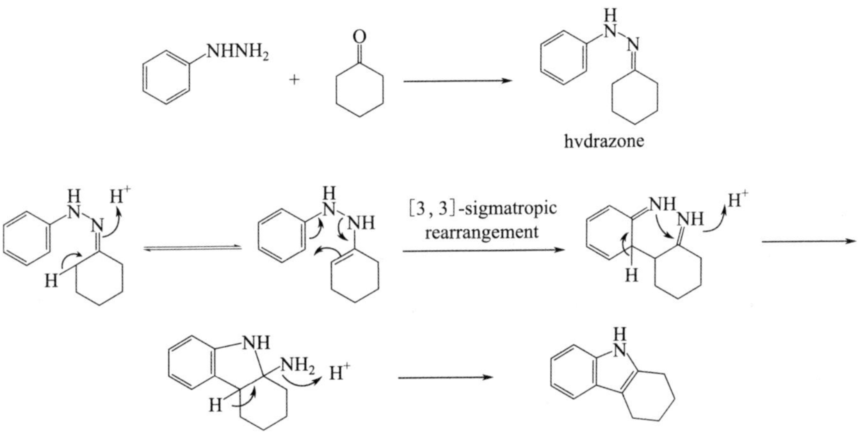

【Apparatus and Reagents】

(1) Apparatus: 100W microwave instrument, flask, condenser pipe.

(2) Reagents: Phenylhydrazine, cyclohexanone, acetic acid, methanol.

【Experimental Procedure】

(1) Conventional heating

A mixture of 0.49g (0.005mol) of cyclohexanone and 10mL of acetic acid contained in a three-necked round-bottom flask equipped with a reflux condenser, a slip-sealed stirrer, and a dropping funnel is heated under reflux and stirred while 0.36g (0.003mol) of phenylhydrazine is added during 1h. The mixture is heated under reflux for an additional hour and poured into a beaker and stirred by hand while it solidifies. It is then cooled to about 5℃ and filtered with suction, the filtrate being cooled in ice and refiltered through the filter cake. The final filtrate is discarded. The filter cake is washed with 50mL of water. Each wash is allowed to soak into the filter cake before it is sucked dry. The crude solid is air-dried overnight and crystallized from methanol after treatment with carbon, then yield 76%—85% of 1,2,3,4-tetrahydrocarbazol, m.p. 115—116℃.

(2) Microwave reaction

A mixture of 0.49g (0.005mol) of cyclohexanone and 10mL of acetic acid was contained in a round bottom flask equipped with a stirred bar. To this solution was slowly added 0.36g (0.003mol) of phenylhydrazine. The reaction exhibits self-heating and caking. The resulting hydrazone was ground into a homogeneous mass and set into microwave reaction. The reaction was carried out in an open vessel and temperature-controlled at 120℃, power: 100W, run time: 1min using a commercial microwave instrument (CEM). While reaction was done, the previously methods was used to get the product.

3.23 金属有机化合物

实验74 二茂铁的合成

【目的与要求】

(1) 掌握二茂铁的合成原理。

(2) 巩固回流、重结晶等基本操作。

【原理】

二茂铁是环戊二烯与 Fe^{2+} 的络合物，环戊二烯负离子的 π 电子数为 6，符合 Hückle 规则，具有芳香性，能发生磺化、烷基化、酰基化等反应。二茂铁及其衍生物可用作紫外吸收剂，火箭燃料的添加剂。

$$\frac{2}{3}FeCl_3 + \frac{1}{3}Fe \longrightarrow FeCl_2$$

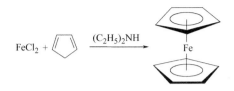

环戊二烯负离子的形成方法有多种，除了上述有机碱法，还有如下几种。

碱金属氢氧化物法：

$$C_5H_6 + KOH \longrightarrow C_5H_5K + H_2O$$

金属钠法：

$$C_5H_6 + Na \longrightarrow C_5H_5Na + \frac{1}{2}H_2$$

格氏试剂法：

$$C_5H_6 + RMgX \longrightarrow C_5H_5MgX + RH$$

【仪器与药品】

(1) 仪器：25mL 三口烧瓶，分液漏斗，滴液漏斗，H 形分馏管，温度计，电磁搅拌器。

(2) 药品：无水三氯化铁，还原铁粉，环戊二烯，二乙胺，四氢呋喃，石油醚，乙醇，氮气。

【步骤】

在 25mL 三口烧瓶上安装 H 形分馏管、温度计及氮气导管。加入 12mL 四氢呋喃，通入氮气，开动电磁搅拌器，分别加入 2.7g 无水三氯化铁和 0.5g 还原铁粉，加热回流 3h，得到带有灰色沉淀物的棕色液体。减压蒸馏除去溶剂，得到二氯化铁。反应瓶用水冷却，并继续通入氮气，用滴液漏斗慢慢滴加 10mL 二乙胺和 4.2mL 环戊二烯的混合溶液，控制反应温度约在 20℃[1]。继续搅拌 6h，蒸馏除去二乙胺，加入石油醚 15mL，回流，过滤，滤渣重复提取 2 次，合并石油醚提取液，蒸馏除去石油醚，得粗产物。加入乙醇重结晶[2]，得橘黄色针状结晶，其熔点为 173～174℃。二茂铁的红外光谱见图 3-11，其核磁共振谱见图 3-12。

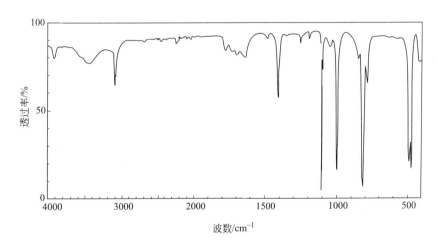

图 3-11 二茂铁的红外光谱图

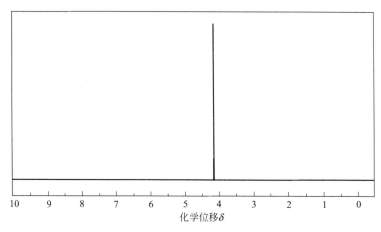

图 3-12 二茂铁的核磁共振谱图（溶剂：氘代氯仿）

【附注】

[1] 反应是放热反应。

[2] 也可用升华法提纯。

实验 75 丁基锂的合成

【目的与要求】

(1) 掌握丁基锂的合成原理和合成方法；

(2) 巩固无水、无氧操作。

【原理】

丁基锂是非常重要的金属有机化合物，可用于合成其他金属有机化合物，广泛应用于合成医药、农药、有机高分子化合物等。其合成主要是通过金属锂与氯丁烷反应。

$$CH_3CH_2CH_2CH_2Cl + 2Li \longrightarrow CH_3CH_2CH_2CH_2Li + LiCl$$

丁基锂性质活泼，纯晶遇水或潮湿空气反应猛烈，可着火，容易与活泼氢和氧气等反应。因此，丁基锂在制备和使用时需要保持无水和无氧。丁基锂在烃类溶剂中比较稳定，但洒在可燃物上也会导致着火，易燃程度视所用溶剂而定。

【仪器与药品】

(1) 仪器：25mL 圆底烧瓶，分液漏斗。

(2) 药品：正己烷，金属锂片，氯丁烷，氩气，2,2′-联二吡啶，二甲苯，仲丁醇。

【步骤】

(1) 丁基锂的制备 在氩气保护下，于 125mL 三口烧瓶中加入 75mL 正己烷（或 60～90 石油醚），将 3.5g（0.5mol）金属锂片用正己烷（或 60～90 石油醚）洗涤干净，戴上一次性手套，将金属锂片快速撕成小片，加入到三口瓶中，装上机械搅拌，冰盐浴冷却至 0℃ 左右（**注意：温度别太低，否则引发比较慢**），往其中滴加 23.2g（0.25mol）氯丁烷，控温在 15℃ 以下（**注意：反应引发后为紫灰色，开始时应该慢慢滴加，反应放热比较剧烈，特别注意别冲料**），加完后，冰盐浴控温 15℃ 以下继续搅拌 2h，然后撤去冰盐浴，室温搅拌 1h，然后改为回流装置，逐渐升温回流 4～5h，冷却至室温，静置沉降过夜，上清液为丁基锂溶液，用氩气压至储存瓶中，残渣加入50mL溶剂搅拌，沉降过夜，上清液合并到丁基锂溶液中备用。

（2）残渣处理　将上面残渣带瓶放在一个装有沙土的盆中，置于安全的地方，暴露在空气中1~2个月（不会自燃），待体系中固体变为白色时，立刻倒出来，在空地上远远的往上面泼水即可。如果当地气候比较潮湿时，开始几天应在两边的瓶口上加上干燥管。如果急用反应瓶，在转移出丁基锂溶液后，可以在氩气保护下，冰盐浴冷却，缓慢的滴加叔丁醇处理。

（3）标定方法　丁基锂在用前必须标定，有双滴定法和单滴定法。双滴定法比较烦琐，一般可用单滴定法滴定。具体方法如下。

① 滴定试剂：1mol/L的仲丁醇/二甲苯溶液（仲丁醇和二甲苯均须用活化过的5A分子筛干燥）。

② 指示剂：2,2′-联吡啶。

③ 溶剂：二甲苯（须用活化过的5A分子筛干燥）。

④ 操作方法：在氩气保护下，往带一个翻口塞的100mL的三口瓶中加入磁子，20mL二甲苯，很少量指示剂，然后用一个带精确刻度的2mL注射器准确量取2mL丁基锂快速转移到瓶中（注射器中的空气需要用氩气置换，且抽取丁基锂时需排除针筒中的氩气，并在丁基锂溶液中来回抽排几次，以免针筒中微量的水和空气影响滴定的准确度），体系变为紫红色，然后将同一针筒（用同一个针筒的目的是为了减小滴定误差）洗干净，吹干，用滴定剂洗涤两至三次，准确量取滴定剂滴定之体系突变为黄色，即为滴定终点。

⑤ 重复滴定一次，两次误差在2%以内，则可认为结果准确。

⑥ 滴定结果：滴定剂量（mL）/2即为丁基锂的物质的量浓度。

【附注】

[1] 丁基锂遇空气极易自燃，量取时，针尖处在空气中会冒火星。

[2] 整个过程中须用氩气保护，特别需要注意安全。

[3] 丁基锂着火时，须用沙土灭火。平时须在伸手可及的地方备有灭火的沙土。

[4] 制备和使用丁基锂时，最好不要一个人单独操作，以免意外情况时，一个人无法处理。

Experiment 76　Synthesis of Acetyl Ferrocene

【Experimental Objectives】

(1) Understand the principle of Friedel-Crafts acylation.

(2) Learn and master the operations of thin layer chromatography and column chromatography.

【Experimental Principles】

Ferrocene is an example of an organometallic compound, i. e. it contains carbon-metal bonds. In the case of ferrocene, all ten carbon atoms are bonded to iron through the π-electrons. This kind of compound is sometimes referred to as a sandwich compound.

Ferrocene is very stable. It possesses some aromatic character and will undergo many of the typical reactions of purely organic aromatic compounds, such as Friedel-Crafts acylation. Depending on the conditions, Friedel-Crafts acetylation of ferrocene can give a mixture of unreacted ferrocene, acetyl ferrocene and diacetyl ferrocene in which both rings have been acylated. In this experiment, the mixture produced will be analysed by TLC and separated by column chromatography.

$$\text{Ferrocene} \xrightarrow[H_3PO_4]{Ac_2O} \text{Fe-COCH}_3 \xrightarrow[H_3PO_4]{Ac_2O} \text{(COCH}_3)_2\text{Fe}$$

【Apparatus and Reagents】

(1) Apparatus: Heating mantle, round bottom flask, beaker, spherical condenser, thermometer, Buchner funnel, suction flask, Pasteur pipet, TLC plates, chromatographic column, rotary evaporator.

(2) Reagents: Ferrocene, acetic anhydride, phosphoric acid, sodium hydroxide, hexane, ethyl acetate, toluene, silica gel.

【Experimental Procedure】

Set up a water bath at 70℃. Watch the thermometer carefully, to keep the temperature between 70 and 80℃. It is important that the temperature is not below 70℃, otherwise there may be no reaction. Ferrocene is a very stable compound and will survive higher temperatures, so it is better to upwards, rather than downwards.

Place ferrocene (0.25g)[1], acetic anhydride (2.5mL) and 85% phosphoric acid[2] (0.5mL) in a 25mL round bottom flask fitted with a condenser (cooling water is not required) and a drying tube[3]. Heat the mixture in the water bath at 60℃ with stirring for 30min, then pour into ice water (10mL). Neutralize the mixture with NaOH (aq 2mol/L). Filter off the product. Analyze the crude product by TLC (Silica gel, hexane: ethyl acetate=8:1). From the top to bottom, the yellow, orange and red spots on TLC plate represent ferrocene, acetyl ferrocene and 1,1'-diacetyl ferrocene, respectively. Then determine the R_f values.

Dissolve the precipitated reaction product in 3-5mL of toluene and carefully apply it to the top of the column using a long Pasteur pipet. Allow the toluene to run into the silica.

Using a long Pasteur pipet at first, and then by careful pouring, add 100mL of hexane to the top of the column. Apply pressure to the top so that the solvent flows through at a reasonable rate. Collect the solvent in test tubes (so that you know the sequence of tubes) and observe what happens on the column as the solvent flows down. After running all the solvent, check the fractions by TLC. Compare them to authentic ferrocene to see which band (if either) belongs to unreacted starting material. If the first orange fraction has completely come off the column, refill with 4:1 hexane: ethyl acetate and continue to elute a second colored band. Transfer tubes containing identical components into pre-weighed round bottom flasks, taking care to keep the flasks less than half full. Evaporate the solvents on the rotary evaporator. Record the weights of both compounds and calculate the yields.

【Notes】

[1] Weigh the ferrocene directly into the flask. The two liquids can be added by syringe.

〔2〕When adding phosphoric acid, make sure to add it slowly under shake.

〔3〕The flask should be dried, and the drying tube should be used to avoid the water from air into the flask.

【Thinking Questions】

(1) What is the purpose of adding sulfuric acid in the preparation of ethyl acetate? Would any of the ester be formed in the absence of sulfuric acid?

(2) What procedures may be used to drive an esterification toward completion?

(3) Why the reaction mixture would be washed by saturated aqueous Na_2CO_3 solution, saturated aqueous NaCl solution, saturated aqueous $CaCl_2$ solution? Can we wash the reaction mixture with saturated aqueous $CaCl_2$ solution first? Why?

3.24 药物的合成

实验77 局部麻醉剂苯佐卡因

【目的与要求】

(1) 了解局部麻醉剂的有关知识。

(2) 掌握苯佐卡因的合成方法。

【原理】

局部麻醉剂或称止痛剂（Painkiller）是一组已被研究得很透彻的化合物。最早的麻醉剂是从古柯中提取的生物碱可卡因（结构如下所示），虽然其有较好的麻醉作用，但是如果用量大时便成了一种成瘾药品。人们一直在寻找一种理想的局部麻醉剂，现已发现有活性的这类药物均具有共同的结构特征。

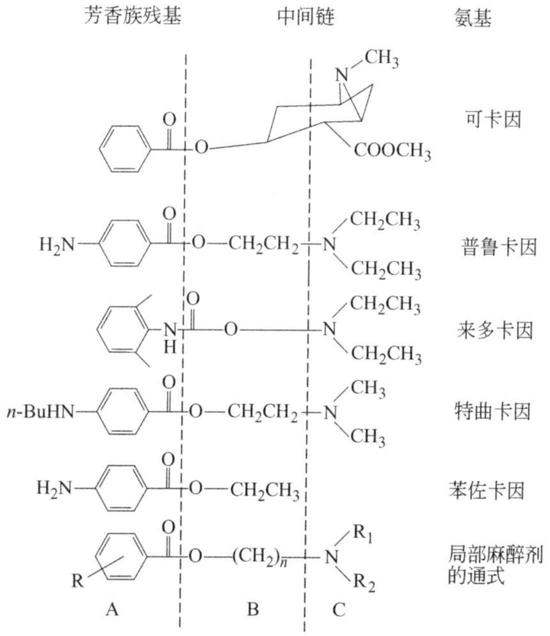

苯佐卡因（Benzocaine）是一有效的局部麻醉剂，主要供皮肤擦剂之用。其合成方法如下：

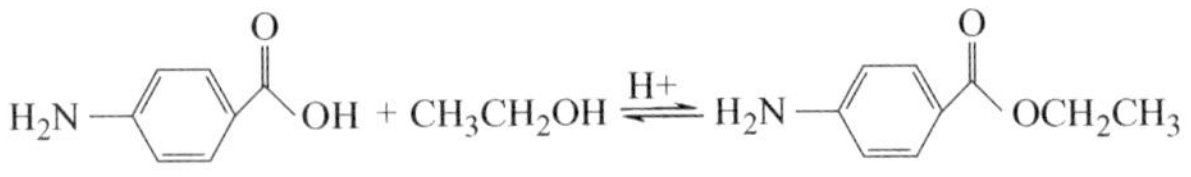

$$H_2N-C_6H_4-COOH + CH_3CH_2OH \xrightleftharpoons{H^+} H_2N-C_6H_4-COOCH_2CH_3$$

【仪器与药品】

(1) 仪器：25mL 圆底烧瓶，分液漏斗。
(2) 药品：对氨基苯甲酸，浓硫酸，乙醚，碳酸钠，95％乙醇。

【步骤】

向 25mL 圆底烧瓶中加入 4g 对氨基苯甲酸，12mL 95％乙醇和 1mL 浓硫酸，将混合物回流 1h，冷却，用 10％的碳酸钠溶液中和，用乙醚萃取，水浴上蒸去乙醚，残余固体用乙醇-水进行重结晶，真空抽滤。苯佐卡因为白色晶体，熔点 89～92℃。

实验 78　磺胺药物——磺胺

【目的与要求】

(1) 了解磺胺类药物的有关知识。
(2) 掌握磺胺的合成方法。

【原理】

1932 年，德国的染料制造商 I. G. Farbenindustrie 获得了红色偶氮染料百浪多息（Prontosil）的专利权，在使用时却意外发现它有抗菌作用而得到了广泛应用。但是进一步的研究发现百浪多息只有在活的动物体内才有效，而在体外不显示药物活性。研究表明百浪多息在体内被代谢成磺胺，磺胺不管在体内还是在体外均有活性，有关发现引起了对磺胺衍生物的兴趣，数年之内被合成的磺胺衍生物达 1 千多种，形成了第一种有用的抗菌药物——磺胺药物。

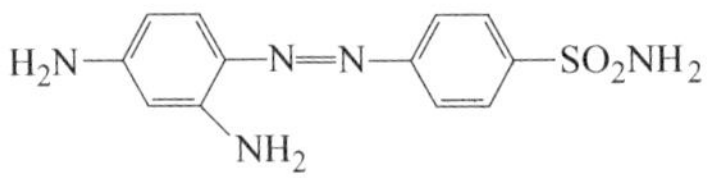

百浪多息 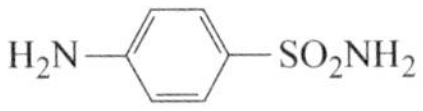　　　　　　　　　　　磺胺

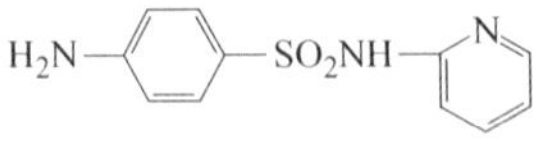

磺胺吡啶 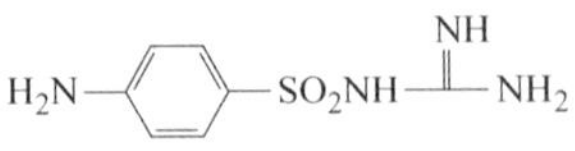　　　　　　　　　　　磺胺胍

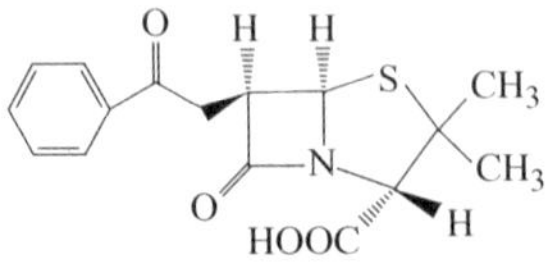

青霉素 G

虽然如青霉素等抗生素的大量生产，使得磺胺药物开始失去其作为普遍使用的抗生剂的重要性，但是磺胺在治疗诸如疟疾、肺结核、麻风病、脑膜炎、猩红热、鼠疫、呼吸道感染、肠内或尿路感染等疾病方面仍然有其广泛的用途。

磺胺的合成方法如下：

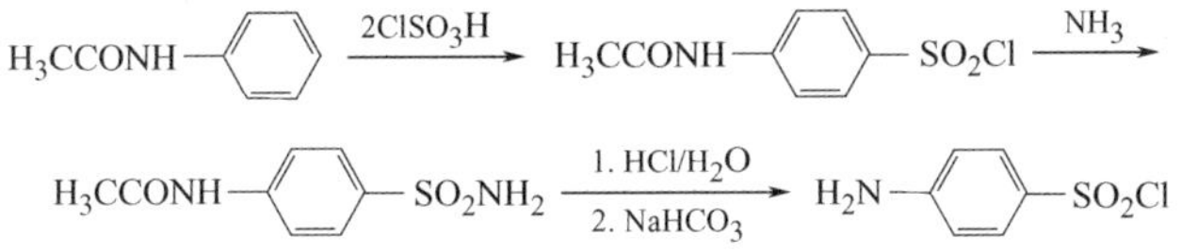

$$H_3CCONH-\bigcirc \xrightarrow{2ClSO_3H} H_3CCONH-\bigcirc-SO_2Cl \xrightarrow{NH_3}$$

$$H_3CCONH-\bigcirc-SO_2NH_2 \xrightarrow[2.\ NaHCO_3]{1.\ HCl/H_2O} H_2N-\bigcirc-SO_2Cl$$

【仪器与药品】

(1) 仪器：25mL 锥形瓶，分液漏斗。

(2) 药品：乙酰苯胺，氯磺酸，浓氨水，碳酸氢钠，浓盐酸。

【步骤】

(1) 对乙酰氨基苯磺酰胺的合成　25mL 锥形瓶中加入 3g 乙酰苯胺，加热熔化，除去表面吸附的水汽，同时旋摇锥形瓶，使其均匀沉积于瓶底和瓶壁。置锥形瓶于冰水浴中冷却，将 9mL 氯磺酸 1 次加入烧瓶中，并立即将装有导气管的橡皮塞塞上，然后旋摇烧瓶，如果反应过于剧烈，需要稍加冷却，反应缓慢时，将烧瓶置于水浴中加热 10min[1]。撤去吸收装置，将烧瓶置于通风橱内的冰水浴中冷却。当充分冷却后，将反应液慢慢倒入 40g 冰和 40g 水的混合物中，并用冰水洗涤烧瓶，过滤收集固体产物，研碎并用冰水洗涤[2]。抽滤，得对乙酰氨基苯磺酰氯。将上述粗对乙酰氨基苯磺酰氯放入小烧杯中，边搅拌、边慢慢滴加 12mL 浓氨水，产生白色糊状物。滴加完毕后，继续搅拌 15min，再放入 60℃ 的水浴中搅拌 15min，室温放置冷却后，抽滤，用 5mL 冷水洗涤，抽干，得对乙酰氨基苯磺酰胺。

(2) 磺胺（对氨基苯磺酰胺）的合成　将所得到的对乙酰氨基苯磺酰胺、3mL 浓盐酸和 10mL 水及沸石加入到 30mL 圆底烧瓶中，缓慢回流 0.5h，得黄色近透明的溶液，再加入 5mL 水及 0.2g 活性炭，煮沸，抽滤。滤液倒入烧杯中，在搅拌下加入粉状碳酸氢钠使溶液呈碱性[3]。冰水浴冷却使沉淀完全，抽滤，用少量的冰水洗涤。粗产品可用乙醇重结晶。纯品磺胺的 m.p. 为 163~164℃。

【附注】

[1] 此过程注意防止倒吸。

[2] 因为酰氯容易与水反应，有关过程尽量要快。

[3] 产生的二氧化碳泡沫可能会使反应物溢出烧杯，可通过控制加入碳酸氢钠的速度和加快搅拌的速度来防止。

实验 79　解热镇痛药——非那西汀

【目的与要求】

(1) 了解解热镇痛药非那西汀的有关知识。

(2) 掌握非那西汀的合成方法。

【原理】

非那西汀（Phenacetin，对乙酰氨基苯乙醚）有显著的解热镇痛效果。本品口服毒性低，长期以来曾广泛用于治疗关节痛、神经痛、头痛、发热感冒等，也是复方阿司匹林（A.P.C.，Aspirin、Phenacetin 和 Caffeine）的一个成分。然而，大剂量服用后可产生过量的变性血红蛋白，导致眩晕、发绀、呼吸困难等副作用，特别是长期大剂量服用，会对肾脏、血红蛋白及视网膜产生损害。1982年9月，我国卫生部发出了淘汰127种药品的通知，其中包括非那西汀片剂和含非那西汀的小儿退热片，但仍保留了非那西汀原料和含非那西汀的复方制剂，如A.P.C.等。

非那西汀的合成在有机合成中具有一定的意义，它将一些基本而又重要的实验糅合在一起，因而值得掌握。其合成的路线有好几条，主要有：

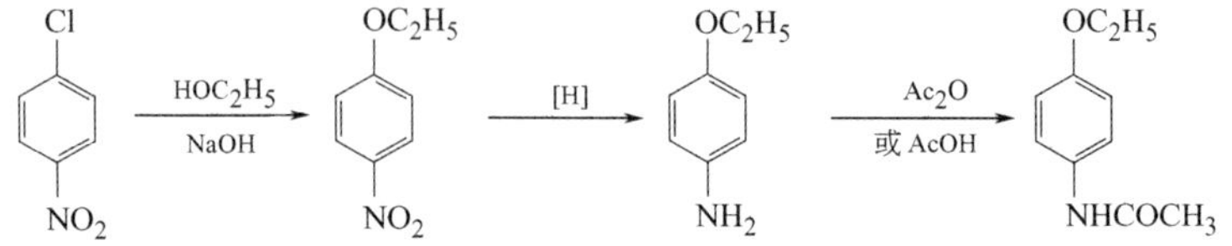

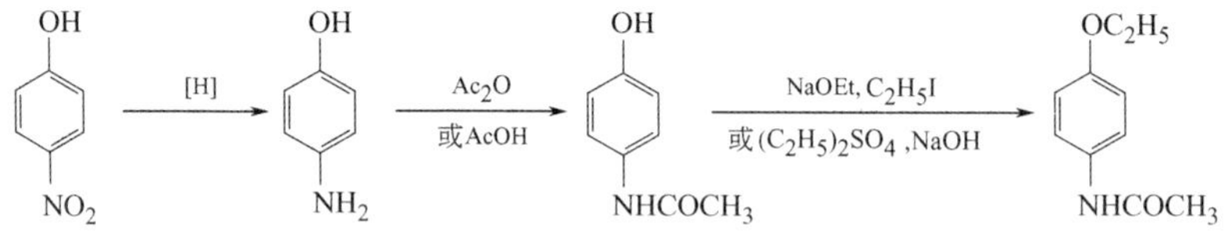

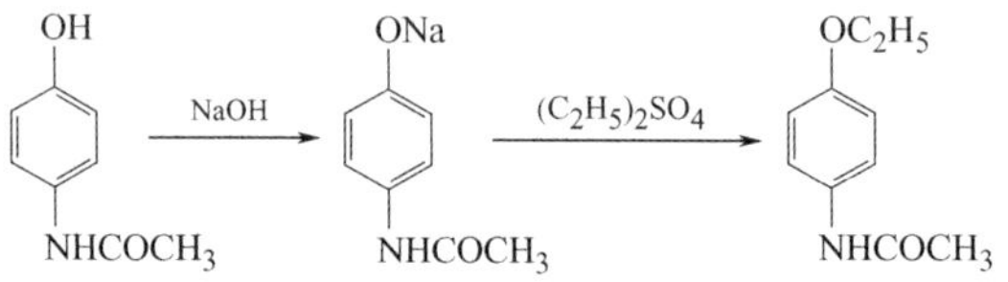

本实验采用第一种方法，该方法简单，易操作。

【仪器与药品】

(1) 仪器：30mL锥形瓶，分液漏斗，磁力加热搅拌器。

(2) 药品：对乙氧基苯胺，乙酸酐，$CH_3COONa \cdot 3H_2O$，浓盐酸。

【步骤】

在30mL锥形瓶中加入18mL水、1g对乙氧基苯胺和浓盐酸0.7mL，将胺完全溶解。置于蒸汽浴上加热，并开动磁力加热搅拌器，在搅拌下加入0.9mL乙酸酐，立即加入乙酸钠缓冲溶液（1.2g $CH_3COONa \cdot 3H_2O$ + 4mL H_2O）。剧烈搅拌，使溶液充分混合。

将烧瓶置于冰水浴中冷却，继续搅拌至非那西汀结晶析出。真空过滤，冰水洗涤，得粗产品。产品可以用水重结晶。非那西汀为白色有闪光的鳞片状结晶，熔点134～136℃。

实验80 止 咳 酮

【目的与要求】

(1) 了解合成止咳酮的原理和方法。

(2) 了解乙酰乙酸乙酯在有机合成中的应用以及酮与亚硫酸氢钠的反应。

(3) 巩固洗涤、分液、干燥、蒸馏和抽滤等基本操作。

【原理】

4-苯基-2-丁酮与亚硫酸氢钠的加成产物一般称为止咳酮，可用于止咳药物的制剂与保存。其合成一般首先由3-丁酮酸乙酯（乙酰乙酸乙酯）合成得到4-苯基-2-丁酮，然后再与亚硫酸氢钠反应（反应式如下所示）。

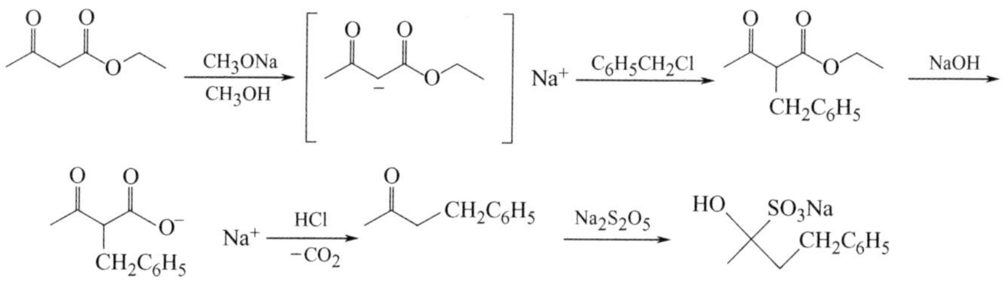

【仪器与药品】

(1) 仪器：30mL 三口烧瓶，分液漏斗，滴液漏斗，砂芯漏斗，圆底烧瓶，微型蒸馏头，球形冷凝管，电磁加热搅拌器，温度计。

(2) 药品：乙酰乙酸乙酯，氯化苄，28%甲醇钠，焦亚硫酸钠，氢氧化钠，浓盐酸，无水硫酸镁，95%乙醇。

【步骤】

(1) 4-苯基-2-丁酮的合成 在装有滴液漏斗、温度计和球形冷凝管的 30mL 三口烧瓶中加入 8mL 甲醇钠溶液，室温搅拌下滴加 6mL 乙酰乙酸乙酯，滴加完毕搅拌 10min。随后继续滴加 4.5mL 氯化苄，约 15min 滴完，继续搅拌 10min，然后回流 1.5h。稍冷，慢慢在 5~10min 内加入 2g 氢氧化钠与 32mL 水配成的溶液，加热回流 1.5h，有油层析出。停止加热并冷却到 40℃以下，缓慢加入浓盐酸 5mL 至溶液变黄（pH=1~2），再回流 1.5h。改蒸馏装置，蒸馏除去低沸物，残余物分出有机层，得粗产物（可直接用于下一步反应）。提纯方法为在无水硫酸镁干燥后，减压蒸馏，收集 111℃/1.596kPa 的馏分。

(2) 亚硫酸氢钠加成物的合成 将上述产品加入 95%乙醇，并加热至 60℃，得溶液 A；将 1.8g 焦亚硫酸钠与水 9mL 加热至 80℃，得透明溶液 B。趁热将溶液 A 慢慢倒入溶液 B 中，并不时搅拌，然后回流 15min，溶液透明。冷却，待结晶完全后，过滤，用少量 95%的乙醇洗涤，晾干。

实验 81　抗癫痫药 5,5-二苯基乙内酰脲

【目的与要求】

(1) 掌握二苯乙二酮和尿素的缩合反应原理及实验操作。

(2) 熟练回流、重结晶等操作。

【原理】

5,5-二苯基乙内酰脲即苯妥英，英文名 Phenytoin，是一种抗癫痫药。属Ⅰb类抗心律失常药，具有膜稳定性，抑制快钠离子内流，适用于治疗全身性强直阵挛性发作、复杂部分性发作（精神运动性发作、颞叶癫痫）、单纯部分性发作（局限性发作）和癫痫持续状态。也可用于治疗三叉神经痛。

二苯基乙二酮是一个不能烯醇化的 α-二酮，在碱的作用下发生重排，生成二苯基乙醇酸。此反应是由羟基负离子向二苯基乙二酮分子中的羰基加成，形成活性中间体而开始的。此时另一个羰基则是亲电中心苯基带着一对电子进行转移重排，反应的动力是所生成的羧基负离子的稳定性。生成的二苯基乙醇酸再与尿素缩合，得到 5,5-二苯基乙内酰脲。该化合物的合成是一个多步骤反应。

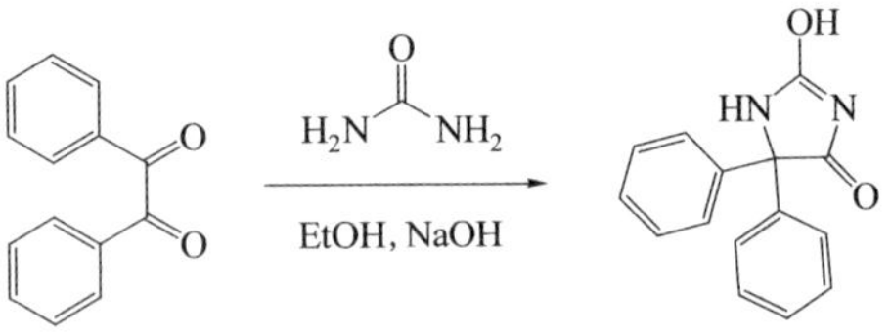

$$\underset{\text{EtOH, NaOH}}{\overset{H_2N\underset{\overset{\parallel}{O}}{-}NH_2}{\longrightarrow}}$$

【仪器与药品】

（1）仪器：圆底烧瓶，电热套，球形冷凝管，滴液漏斗，分液漏斗，布氏漏斗，吸滤瓶。

（2）药品：二苯乙二酮，尿素，10%盐酸，30%氢氧化钠，95%乙醇。

【步骤】

在25mL圆底烧瓶中加0.5g二苯乙二酮、0.25g尿素、8mL 95%乙醇和1.5mL 30%氢氧化钠水溶液，水浴加热回流1.5h。水浴冷却，将反应混合物倒入盛有15mL水的烧杯中，抽滤，除去固体杂质。向滤液中滴加10%盐酸，直至石蕊试纸呈酸性反应（pH=5~6）为止。抽滤，收集固体为5,5-二苯基乙内酰脲粗产品，用10mL水洗一次。用95%乙醇重结晶，加活性炭脱色，晾干，称量，计算产率。

思 考 题

（1）请给出5,5-二苯基乙内酰脲的IUPAC系统名称。

（2）请给出确证本产物结构的方法。

Experiment 82　Synthesis of Aspirin (acetylsalicylic acid)

【Experimental Objectives】

(1) Understand the acetylation process.

(2) Master the recrystallization operation process.

【Experimental Principles】

Analgesics are compounds used to reduce pain, antipyretics are compounds used to reduce fever. One popular drug that does both is aspirin (acetylsalicylic acid). The Merck Index, which is an enclopedia of chemicals, drugs and biologicals, lists the following information under aspirin: acetylsalicylic acid, monoclinic tablets or needle-like crystals, m. p. 135 (rapid heating), odorless, but in moist air it is gradually hydrolyzed into salicylic and acetic acids, one gram dissolves in 300mL of water at 25℃, in 100mL of water at 37℃, in 5mL alcohol, in 17mL chloroform.

Phenols, unlike amines, cannot be acetylated satisfactorily in aqueous solution, acetylation proceeds readily with acetic anhydride in the presence of a little concentrated sulphuric acid as catalyst.

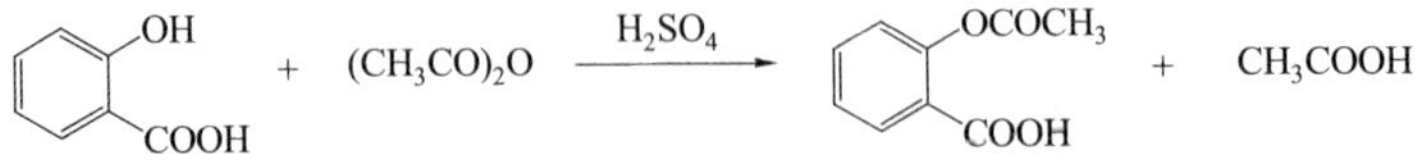

$$\text{salicylic acid} + (CH_3CO)_2O \xrightarrow{H_2SO_4} \text{acetylsalicylic acid} + CH_3COOH$$

【Apparatus and Reagents】

(1) Apparatus: Flask, water bath, thermometer.

(2) Reagents: Salicylic acid, acetic anhydride, concentrated sulphuric acid, acetic acid, water.

【Experimental Procedure】

Place 1.0g of dry salicylic acid and 1.5g (1.4mL) of acetic anhydride in a small conical flask, add 1drops of concentrated sulphuric acid, and rotate the flask in order to secure thorough mixing. Warm on a water bath to about 50—60℃, stirring with the thermometer, for about 15min. Allow the mixture to cool and stir occasionally. Add 15mL of water, stir well and filter at the pump. Recrystallize the crude acetylsalicylic acid from a mixture of equal volumes of acetic acid and water.

The following is an alternative method of purifying the crude aspirin. Dissolve the solid in about 3mL of hot alcohol and pour the solution into about 7.5mL of warm water; if a solid separates at this point, warm the mixture until solution is complete and then allow the clear solution to cool slowly. Beautiful needle-like crystals will separate. The yield is 1.3g. The air-dried crude product may also be recrystallised from benzene or from ether-light petroleum (b.p. 40—60℃).

【Notes】

Acetylsalicylic acid decomposes when heated and does not possess a true, clearly-defined m.p. Decomposition points ranging from 128℃ to 135℃ have been recorded. Some decomposition may occur if the compound is recrystallized from a solvent of high boiling point or if the boiling period during recrystallization is unduly prolonged.

【Ferric Chloride Test for Salicylic Acid】

Add 10 drops of aqueous 1% ferric chloride solution to a test tube containing a few crystals of the compound to be tested dissolved in 5mL water and note the color. Do this test with phenol, salicylic acid, and your crude product. Formation of an iron-phenol complex with Fe(Ⅲ) gives a definite color ranging from red to violet, depending upon the particular phenol present.

【Questions】

(1) Normally, you measure reactants to at least two significant figures. Why is it not necessary to measure the volume of acetic anhydride to two significant figures? What is the theoretical yield of aspirin (in two significant figures).

(2) Why is the aspirin washed with cold water?

(3) According to the data in the Merck Index, if 1.0g of aspirin is dissolved in 100mL of water at 37℃, how much aspirin will precipitate out of solution when it is cooled to 25℃?

(4) A polymeric material, which is a polyester, is formed in this reaction. Polyesters are often made from dicarboxylic acids and diols. In this case, one molecule (salicylic acid) provides both the "alcohol" and the carboxylic acid. Write a structure for the ester formed from acetic acid and ethanol. If you have difficulty, look at the banana oil experiment.

3.25 农药的合成

实验 83 杀灭菊酯（氰戊菊酯）

【目的与要求】
(1) 学习杀灭菊酯的合成原理和方法。
(2) 掌握剧毒化学品氰化钠的使用方法和处理方法与注意事项。
(3) 巩固掌握洗涤、分液、干燥、蒸馏等基本操作。

【原理】
杀灭菊酯（又称氰戊菊酯）具有广谱、高效、低毒等特点，适用于农、林、牧、果、蔬菜、烟、茶等作物及家畜害虫的防治。其合成方法如下：

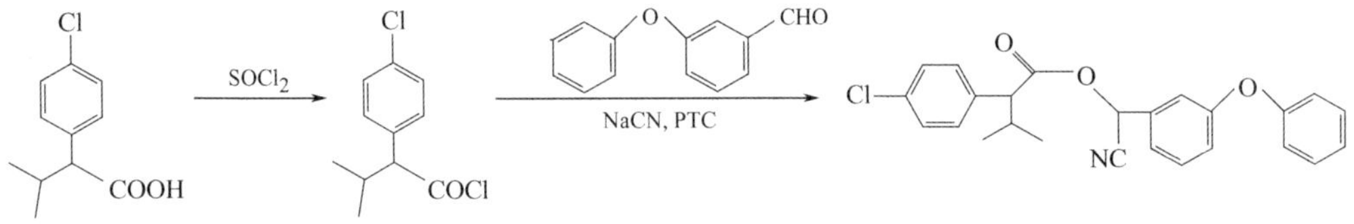

其中第二步首先是间苯氧基苯甲醛（简称醚醛）、氰化钠在相转移催化剂的存在下进行反应得到氰醇，氰醇进一步与酰氯反应得到杀灭菊酯。

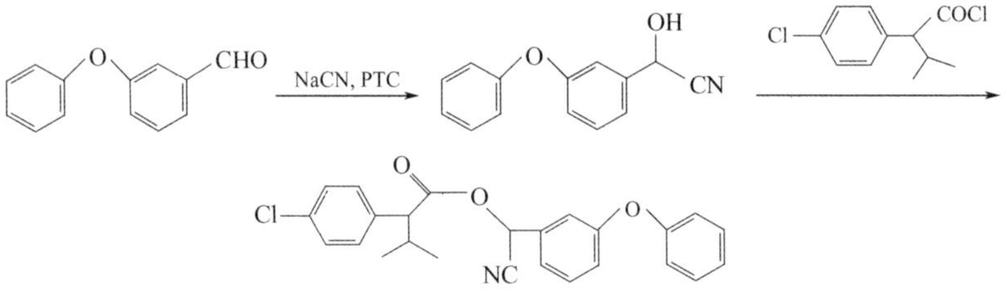

【仪器与药品】
(1) 仪器：30mL 三口烧瓶，分液漏斗，滴液漏斗，砂芯漏斗，圆底烧瓶，微型蒸馏头，球形冷凝管，磁力加热搅拌器，温度计。
(2) 药品：α-异丙基对氯苯乙酸，甲苯，氯化亚砜，氯化四丁铵，氰化钠，正庚烷，醚醛。

【步骤】
(1) α-异丙基对氯苯乙酰氯的合成　在装有滴液漏斗、温度计、球形冷凝管和吸收导管的 30mL 三口烧瓶中加入 4.5g α-异丙基对氯苯乙酸和 10mL 甲苯，在搅拌下滴加 9.5g 氯化亚砜，滴加完毕逐渐升温至 60~80℃，并在此温度下反应 3h。蒸馏除去甲苯和氯化亚砜得 α-异丙基对氯苯乙酰氯。

(2) 杀灭菊酯的合成　在 30mL 圆底烧瓶中加入醚醛 2.0g、α-异丙基对氯苯乙酰氯 2.4g、NaCN[1] 0.5g、水 2mL、氯化四丁铵 0.02g 和正庚烷 20mL。将上述溶液搅拌 5h 后加入 15mL 水搅拌，分去水层[2]，把有机层水洗 2 次，每次 15mL，干燥，蒸馏除去溶剂得粗产物。

【附注】

[1] 氰化钠为剧毒药品，有关操作的注意事项见附录5。
[2] 水层应该用亚硫酸亚铁溶液处理以后才能倒入废液桶中，然后集中处理。

实验84　O,O-二甲基-O-(2,2-二氯乙烯基)磷酸酯——敌敌畏

【目的与要求】

（1）学习由敌百虫合成敌敌畏的原理和方法。
（2）了解分子重排反应。
（3）巩固掌握洗涤、分液、干燥、蒸馏等基本操作。

【原理】

敌百虫的系统名称为 O,O-二甲基-O-(2,2,2-三氯-1-羟基乙基) 磷酸酯，与氢氧化钠溶液反应脱去 HCl，并发生重排反应生成 O,O-二甲基-O-(2,2-二氯乙烯基) 磷酸酯（即敌敌畏，反应式如下所示）。敌敌畏的生物活性比敌百虫的大。

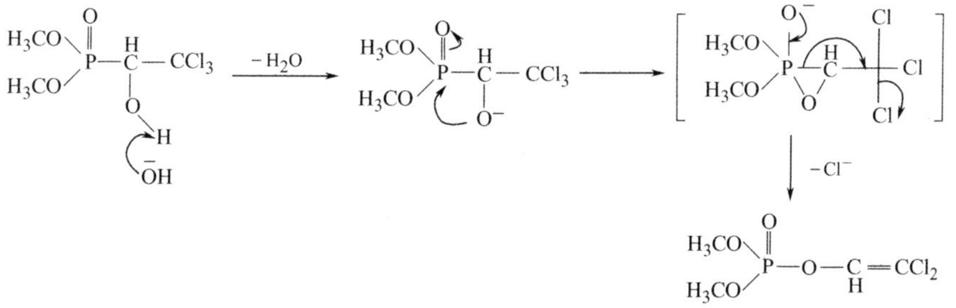

【仪器与药品】

（1）仪器：30mL 三口烧瓶，分液漏斗，滴液漏斗，砂芯漏斗，圆底烧瓶，微型蒸馏头，球形冷凝管，电磁加热搅拌器。

（2）药品：O,O-二甲基-O-(2,2,2-三氯-1-羟基乙基)磷酸酯（敌百虫），苯，5%氢氧化钠溶液，无水硫酸钠。

【步骤】

在装有滴液漏斗、温度计和球形冷凝管的 30mL 三口烧瓶中加入 1.3g 敌百虫和 10mL 热水（50℃），开动磁力加热搅拌器搅拌形成浑浊液，当温度降低到 30℃ 时加入苯 2mL，在搅拌下滴加 5% 氢氧化钠溶液 1.0mL，其间温度保持在 40~50℃，pH 值维持在 9~11。滴加完毕，保温继续反应 10min。冷却至 30℃，分去水层，有机层用等量水洗 1 次后，用无水硫酸钠干燥。蒸馏除去苯，得敌敌畏粗产物。

Experiment 85　Synthesis of Anophelifuge N,N-Diethyl-m-toluamide

【Experimental Objectives】

Master the operation of vacuum distillation.

【Experimental Principles】

N,N-Diethyl-m-toluamide, also called DEET or diethyltoluamide, is the most common active ingredient in insect repellents. It is a slightly yellow oil intended to be applied to the skin or to clothing and provides protection against mosquitoes, ticks, fleas, chiggers, leeches and many biting insects.

It can be prepared by converting *m*-toluic acid (3-methylbenzoic acid) to the corresponding acyl chloride, and allowing it to react with diethylamine.

Here using *m*-toluic acid and diethylamine as raw materials, N,N-diethyl-*m*-toluamide (DEET, b. p. 110℃/1mmHg) was synthesized under the condition of high temperature and normal pressure.

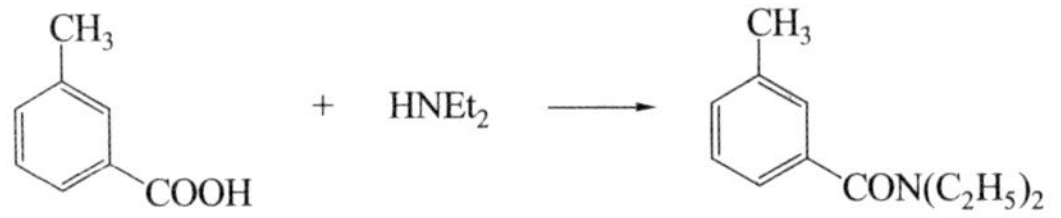

$$\text{m-CH}_3\text{-C}_6\text{H}_4\text{-COOH} + \text{HNEt}_2 \longrightarrow \text{m-CH}_3\text{-C}_6\text{H}_4\text{-CON(C}_2\text{H}_5)_2$$

【Apparatus and Reagents】

(1) Apparatus: Round bottom flask, condenser pipe, heating jacket.

(2) Reagents: *m*-Toluic acid, diethylamine, phosphoric acid.

【Experimental Procedure】

In the 25mL round bottom flask, 0.05mol *m*-toluic acid was added. Phosphoric acid, as a catalysis, was added 1% quality dosage of the *m*-toluic acid. The mixture was heated to dissolve, and 0.15mol diethylamine was added dropwise at 290℃. Then the mixture was keep at this temperature for 1h. After cooling to room temperature, the mixture was washed by 25% NaOH solution, water, and saturated salt solution. The crude product was then through the reduced pressure distillation to collect the distillate at the temperature between 165 and 175℃ to afford pure product.

附 录

附录1 常用有机试剂的物理性质

名 称	熔点/℃	沸点/℃	相对密度 d_4^{20}	折射率 n_D^{20}
甲醇	−98	64.9	0.7914	1.3288
乙醇	−114	78.5	0.7893	1.3611
丁醇	−90.2	117.2	0.8098	1.3993
叔丁醇	25~25.5	83	0.789	1.3878
乙醚	−116.2	34.5	0.741	1.3526
乙酸	16.6	117.9	1.0492	1.3720
乙酐	−73	139.5	1.0820	1.39006
乙酸乙酯	−83.6	77.0	0.9003	1.3723
二氧六环	12	101.7	1.0337	1.4205~1.422
苯	5.5	80.1	0.8786	1.5011
甲苯	−95	110.6	0.8669	1.4961
二甲苯(o-,m-p-)	−25	140.0	0.88	
氯仿	−64	61.7	1.4832	1.4459
四氯化碳	−23	76.5	1.5940	1.4601
二硫化碳	−111	46.2	1.2632	1.6319
硝基苯	5~6	210.8	1.2037	1.5503~1.5523
三正丁基胺	−70	215	0.778	1.4273~1.4293
三乙胺	−115	90	0.7	1.399~1.401
丙酮	−94.9	56.2	0.7899	1.3588
苯乙酮	19.6	202	1.09	1.5321
乙酰水杨酸	135	136~140	1.35	
邻氨基苯甲酸	146	Sub.	1.412	
对氨基苯甲酸乙酯	89~92	172/1.7kPa		
苯胺	−6.2	181~185	1.021	1.5845~1.5865
甲醛	−92	−21	0.815	
正丁醛	−96	75	0.8	1.377~1.383

续表

名　称	熔点/℃	沸点/℃	相对密度 d_4^{20}	折射率 n_D^{20}
苯甲醛	−26	179	1.05	1.544～1.546
二苯甲酮	47～49	305		
溴苯	−31	156	1.49	1.558～1.560
氯化苄	−39	179	1.1	1.537～1.539
叔丁基氯	−25	51～52	0.842	1.3838～1.3858
溴乙烷	−119	38.4	1.460	1.4239
溴丁烷	−112	100～104	1.27	1.4384～1.4404
对硝基氯苯	81～84	242		
环己醇	23.9	161.5	0.962	1.463～1.467
二苯甲醇	65～68	297～298		
氯磺酸	−80	158	1.787	1.437^{14}
二乙胺	−50	55	0.707	1.384～1.386
乙酰乙酸乙酯	−45	180.8	1.0282	1.4199
萘	80.5	218	1.0253	1.5898^{25}
苯乙烯	−31	145～146	0.9060	1.546～1.548
邻苯二甲酰亚胺	238	366		
双酚A	156～158	220/4mmHg		
双酚S	＞245			
3-氨基邻苯二甲酰肼	319～320			

附录2　常用的酸和碱

溶　液	相对密度	质量分数/%	c/(mol/L)	100mL 液体中溶质的质量/g
浓盐酸	1.19	37	12.0	44.0
恒沸点盐酸(252mL浓盐酸+200mL水,恒沸点110 ℃)	1.10	20.2	6.1	22.2
10%盐酸(100mL浓盐酸+321mL水)	1.05	10	2.9	10.5
5%盐酸(50mL浓盐酸+38.05mL水)	1.03	5	1.4	5.2
1mol/L盐酸(41.5mL浓盐酸稀释到500mL)	1.02	3.6	1	3.6
恒沸点氢溴酸(沸点126 ℃)	1.49	47.5	8.8	70.7
恒沸点氢碘酸(沸点127 ℃)	1.7	57	7.6	97
浓硫酸	1.84	96	18	176
10%硫酸(25mL硫酸+398mL水)	1.07	10	1.1	10.7
0.5mol/L硫酸(13.9mL浓硫酸稀释到500mL)	1.03	4.7	0.5	4.9
浓硝酸	1.42	71	16	101
10%氢氧化钠	1.11	10	2.8	11.1
浓氨水	0.9	28.4	15	25.9

附录3 常用酸碱溶液密度组成表

盐 酸

HCl 质量分数/%	相对密度 d_4^{20}	100mL 水溶液中含 HCl 的质量/g	HCl 质量分数/%	相对密度 d_4^{20}	100mL 水溶液中含 HCl 的质量/g
1	1.0032	1.003	22	1.1083	24.38
2	1.0082	2.006	24	1.1187	26.85
4	1.0181	4.007	26	1.1290	29.35
6	1.0279	6.167	28	1.1392	31.90
8	1.0376	8.301	30	1.1492	34.48
10	1.0474	10.47	32	1.1593	37.10
12	1.0574	12.69	34	1.1691	39.75
14	1.0675	14.95	36	1.1789	42.44
16	1.0776	17.24	38	1.1885	45.16
18	1.0878	19.58	40	1.1980	47.92
20	1.0980	21.96			

硫 酸

H_2SO_4 质量分数/%	相对密度 d_4^{20}	100mL 水溶液中含 H_2SO_4 的质量/g	H_2SO_4 质量分数/%	相对密度 d_4^{20}	100mL 水溶液中含 H_2SO_4 的质量/g
1	1.0051	1.005	65	1.5533	101.0
2	1.0118	2.024	70	1.6105	112.7
3	1.0184	3.055	75	1.6692	125.2
4	1.0250	4.100	80	1.7272	138.2
5	1.0317	5.159	85	1.7786	151.2
10	1.0661	10.66	90	1.8144	163.3
15	1.1020	16.53	91	1.8195	165.6
20	1.1394	22.79	92	1.8240	167.8
25	1.1783	29.46	93	1.8279	170.2
30	1.2185	36.56	94	1.8312	172.1
35	1.2599	44.10	95	1.8337	174.2
40	1.3028	52.11	96	1.8355	176.2
45	1.3476	60.64	97	1.8364	178.1
50	1.3951	69.76	98	1.8361	179.9
55	1.4453	79.49	99	1.8342	181.6
60	1.4983	89.90	100	1.8305	183.1

硝 酸

HNO$_3$ 质量分数/%	相对密度 d_4^{20}	100mL 水溶液中含 HNO$_3$ 的质量/g	HNO$_3$ 质量分数/%	相对密度 d_4^{20}	100mL 水溶液中含 HNO$_3$ 的质量/g
1	1.0036	1.004	65	1.3913	90.43
2	1.0091	2.018	70	1.4134	98.94
3	1.01465	3.044	75	1.4337	107.5
4	1.0201	4.080	80	1.4521	116.2
5	1.0256	5.128	85	1.4686	124.8
10	1.0543	10.54	90	1.4826	133.4
15	1.0842	16.26	91	1.4850	135.1
20	1.0050	22.30	92	1.4873	136.8
25	1.1469	28.67	93	1.4892	138.5
30	1.1800	35.40	94	1.4912	140.2
35	1.2140	42.49	95	1.4932	141.9
40	1.2463	49.85	96	1.4952	143.5
45	1.2783	57.52	97	1.4974	145.2
50	1.3100	65.5	98	1.5008	147.1
55	1.3393	73.66	99	1.5056	149.1
60	1.3667	82.00	100	1.5129	151.3

醋 酸

CH$_3$COOH 质量分数/%	相对密度 d_4^{20}	100mL 水溶液中含 CH$_3$COOH 的质量/g	CH$_3$COOH 质量分数/%	相对密度 d_4^{20}	100mL 水溶液中含 CH$_3$COOH 的质量/g
1	0.9996	0.9996	65	1.0666	69.33
2	1.0012	2.002	70	1.0685	74.80
3	1.0025	3.008	75	1.0696	80.22
4	1.0040	4.016	80	1.0700	85.60
5	1.0055	5.028	85	1.0689	90.86
10	1.0125	10.13	90	1.0661	95.95
15	1.0195	15.29	91	1.0652	96.93
20	1.0263	20.53	92	1.0643	97.92
25	1.0326	25.82	93	1.0632	98.88
30	1.0384	31.15	94	1.0619	99.82
35	1.0438	36.53	95	1.0605	100.7
40	1.0488	41.95	96	1.0588	101.6
45	1.0534	47.40	97	1.0570	102.5
50	1.0575	52.88	98	1.0549	103.4
55	1.0611	58.36	99	1.0524	104.2
60	1.0642	63.85	100	1.0498	105.0

氢溴酸

HBr质量分数/%	相对密度 d_4^{20}	100mL水溶液中含HBr的质量/g	HBr质量分数/%	相对密度 d_4^{20}	100mL水溶液中含HBr的质量/g
10	1.0723	10.7	45	1.44465	65.0
20	1.1579	23.2	50	1.5173	75.8
30	1.2580	37.3	55	1.5953	87.8
35	1.3150	46.0	60	1.6787	100.7
40	1.3772	56.1	65	1.7675	114.9

氢碘酸

HI质量分数/%	相对密度 d_4^{20}	100mL水溶液中含HI的质量/g	HI质量分数/%	相对密度 d_4^{20}	100mL水溶液中含HI的质量/g
20.77	1.1578	24.4	56.78	1.6998	96.6
31.77	1.2962	41.2	61.97	1.8218	112.8
42.70	1.4489	61.9			

发烟硫酸

游离SO_3质量分数/%	相对密度 d_4^{20}	100mL水溶液中游离SO_3的质量/g	游离SO_3质量分数/%	相对密度 d_4^{20}	100mL水溶液中游离SO_3的质量/g
1.54	1.860	2.8	10.07	1.900	19.1
2.66	1.865	5.0	10.56	1.905	20.1
4.28	1.870	8.0	11.43	1.910	21.8
5.44	1.875	10.2	13.33	1.915	25.5
6.42	1.880	12.1	15.95	1.920	30.6
7.29	1.885	13.7	18.67	1.925	35.9
8.16	1.890	15.4	21.34	1.930	41.2
9.43	1.895	17.7	25.65	1.935	49.6

氢氧化铵

NH_3质量分数/%	相对密度 d_4^{20}	100mL水溶液中NH_3的质量/g	NH_3质量分数/%	相对密度 d_4^{20}	100mL水溶液中NH_3的质量/g
1	0.9939	9.94	16	0.9362	149.8
2	0.9895	19.79	18	0.9295	167.3
4	0.9811	39.24	20	0.9229	184.6
6	0.9730	58.38	22	0.9164	201.6
8	0.9651	77.21	24	0.9101	218.4
10	0.9575	95.75	26	0.9040	235.0
12	0.9501	114.0	28	0.8980	251.4
14	0.9430	132.0	30	0.8920	267.6

氢氧化钠

NaOH 质量分数/%	相对密度 d_4^{20}	100mL 水溶液中 NaOH 的质量/g	NaOH 质量分数/%	相对密度 d_4^{20}	100mL 水溶液中 NaOH 的质量/g
1	1.0095	1.010	26	1.2848	33.40
2	1.0207	2.041	28	1.3064	36.58
4	1.0428	4.171	30	1.3279	39.84
6	1.0648	6.389	32	1.3490	43.17
8	1.0869	8.695	34	1.3696	46.57
10	1.1089	11.09	36	1.3900	50.04
12	1.1309	13.57	38	1.4101	53.58
14	1.1530	16.14	40	1.4300	57.20
16	1.1751	18.80	42	1.4494	60.87
18	1.1972	21.55	44	1.4685	64.61
20	1.2191	24.38	46	1.4873	68.42
22	1.2411	27.30	48	1.5065	72.31
24	1.2629	30.31	50	1.5253	76.27

氢氧化钾

KOH 质量分数/%	相对密度 d_4^{20}	100mL 水溶液中 KOH 的质量/g	KOH 质量分数/%	相对密度 d_4^{20}	100mL 水溶液中 KOH 的质量/g
1	1.0083	1.008	28	1.2695	35.55
2	1.0175	2.035	30	1.2905	38.72
4	1.0359	4.144	32	1.3117	41.97
6	1.0544	6.326	34	1.3331	45.33
8	1.0730	8.584	36	1.3549	48.78
10	1.0918	10.92	38	1.3765	52.32
12	1.1108	13.33	40	1.3991	55.96
14	1.1299	15.82	42	1.4215	59.70
16	1.1493	19.70	44	1.4443	63.55
18	1.1688	21.04	46	1.4673	67.50
20	1.1884	23.77	48	1.4907	71.55
22	1.2083	26.58	50	1.5143	75.72
24	1.2285	29.48	52	1.5382	79.99
26	1.2489	42.47			

碳 酸 钠

Na$_2$CO$_3$ 质量分数/%	相对密度 d_4^{20}	100mL 水溶液中 Na$_2$CO$_3$ 的质量/g	Na$_2$CO$_3$ 质量分数/%	相对密度 d_4^{20}	100mL 水溶液中 Na$_2$CO$_3$ 的质量/g
1	1.0086	1.009	12	1.1244	13.49
2	1.0190	2.038	14	1.1463	16.05
4	1.0398	4.159	16	1.1682	18.50
6	1.0606	6.364	18	1.1905	21.33
8	1.0816	8.653	20	1.2132	24.26
10	1.1029	11.03			

附录4 常用有机溶剂处理

有机化学反应离不开溶剂，溶剂不仅作为反应介质，在产物的纯化和后处理中也经常使用。市售的有机溶剂有工业级、化学纯和分析纯等各种规格，纯度愈高，价格愈贵。在有机合成中，常常根据反应的特点和要求，选用适当规格的溶剂，以便使反应能够顺利地进行而又符合勤俭节约的原则。某些有机反应（如 Grignard 反应等），对溶剂要求较高，即使微量杂质或水分的存在，也会对反应速率、产率和纯度带来一定的影响。因此掌握有机溶剂的纯化方法，是十分重要的。有机溶剂的纯化，是有机合成工作的一项基本操作。在此介绍了一些实验室中常用的纯化方法。

无 水 乙 醇

市售的无水乙醇一般只能达到 99.5% 的纯度，在许多反应中需用纯度更高的绝对乙醇，经常需自己制备。通常工业用的 95.5% 的乙醇不能直接用蒸馏法制取无水乙醇，因 95.5% 的乙醇和 4.5% 的水形成恒沸点混合物。要把水除去，第一步是加入氧化钙（生石灰）煮沸回流，使乙醇中的水与生石灰作用生成氢氧化钙，然后再将无水乙醇蒸出。这样得到无水乙醇，纯度最高约 99.5%。纯度更高的无水乙醇可用金属镁或金属钠进行处理。

(1) 95.5% 的乙醇初步脱水制取 99.5% 的乙醇

在 250mL 圆底烧瓶中，放入 45g 生石灰、100mL 工业乙醇，装上回流冷凝管，在水浴上回流 2～3h，然后改装为蒸馏装置，进行蒸馏，收集产品 70～80mL。

(2) 用 99.5% 的乙醇制取绝对无水乙醇

用金属镁制取：反应按下式进行

$$2C_2H_5OH + Mg \longrightarrow (C_2H_5O)_2Mg + H_2$$

乙醇中的水分，即与乙醇镁作用形成氧化镁和乙醇。

$$(C_2H_5O)_2Mg + H_2O \longrightarrow 2C_2H_5OH + MgO$$

在 250mL 的圆底烧瓶中，放置 0.8g 干燥纯净的镁条，7～8mL 99.5% 乙醇，装上回流冷凝管，并在冷凝管上端附加一只无水氯化钙干燥管（以上所用仪器都必须是干燥的）。在沸水浴上或用火直接加热使其达微沸，移去热源，立刻加入几滴碘甲烷和几粒碘片（此时注意不要振荡），顷刻即在碘粒附近发生作用，最后可以达到相当剧烈的程度。有时作用太慢则需加热，如果在加碘之后，作用仍不开始，则可再加入数粒碘（一般地讲，乙醇与镁的作用是缓慢的，如所用乙醇含水量超过 0.5% 则作用尤其困难）。待全部镁已经作用完毕后，加入 100mL 99.5% 乙醇和几粒沸石。回流 1h，蒸馏，产物收存于玻璃瓶中，用橡皮塞塞住，这样制备的乙醇纯度超过 99.99%。无水乙醇的沸点为 78.32℃。

无 水 乙 醚

普通乙醚中含有少量水和乙醇，在保存乙醚期间，由于与空气接触和光的照射，除了上述杂质外通常还含有二乙基过氧化物 $(C_2H_5)_2O_2$。这对于要求用无水乙醚作溶剂的反应（如 Grignard 反应）不仅影响反应，且易发生危险。因此，在制备无水乙醚时，首先须检验有无过氧化物存在。为此取少量乙醚与等体积的 2% 碘化钾溶液，再加入几滴稀盐酸一起振摇，振摇后的溶液若使淀粉显蓝色，即证明有过氧化物存在。此时应按下述步骤处理。

在分液漏斗中加入普通乙醚，再加入相当于乙醚体积 1/5 的新配制的硫酸亚铁溶液[1]，剧烈摇动后分去水层。醚层在干燥瓶中用无水氯化钙干燥，间隙振摇，放置 24h，这样可除去大部分水和乙醇。蒸馏[2]，收集 34～35℃ 馏分，在收集瓶中压入钠丝，然后用带有氯化钙干燥管的软木塞塞住，或者在木塞中插入一端拉成毛细管的玻璃管，这样可使产生的气体逸出，并可防止潮气侵入。放置 24h 以上，待乙醚中残留的痕量水和乙醇转化为氢氧化钠和乙醇钠后，才能使用。

【附注】

[1] 硫酸亚铁溶液的配制：在 55mL 水中加入 3mL 浓硫酸，然后加入 30g 硫酸亚铁。此溶液必须在使用时配制，放置过久易氧化变质。

[2] 乙醚沸点低，极易挥发，严禁用明火加热，可用事先准备好的热水浴加热，或者用变压器调节的电热锅加热。尾气出口通入水槽，以免乙醚蒸气散发到空气中。由于乙醚蒸气比空气重（约为空气的 2.5 倍），容易聚集在桌面附近或低洼处。当空气中含有 1.85%～36.5% 的乙醚蒸气时，遇火即会发生燃烧、爆炸，因此在蒸馏过程中必须严格遵守操作规程。

无 水 甲 醇

市售的甲醇，系由合成而来，含水量不超过 0.5%～1%。由于甲醇和水不能形成共沸物，为此可用高效的精馏柱将少量水除去。精制甲醇含有 0.02% 的丙酮和 0.1% 的水，一般已可应用。如要制得无水甲醇，可用镁精制的方法（见无水乙醇）。若含水量低于 0.1%，亦可用 3A 或 4A 型分子筛干燥。甲醇有毒，处理时应避免吸入其蒸气。

无水无噻吩苯

普通苯含有少量的水（可达 0.02%），由煤焦油加工得来的苯还含有少量噻吩（沸点为 84℃），不能用分馏或分步结晶等方法分离除去。为制得无水、无噻吩的苯可采用下列方法。

在分液漏斗内将普通苯及相当苯体积 15% 的浓硫酸一起摇荡，摇荡后将混合物静置，弃去底层的酸液，再加入新的浓硫酸，这样重复操作直至酸层呈现无色或淡黄色，且检验无噻吩为止。分去酸层，苯层依次用水、10% 碳酸钠溶液、水洗涤，用氯化钙干燥，蒸馏，收集 80℃ 的馏分。若要高度干燥可加入钠丝进一步去水。由石油加工得来的苯一般可省去除噻吩的步骤。

噻吩的检验：取5滴苯于小试管中，加入5滴浓硫酸及1～2滴1% α,β-吲哚醌-浓硫酸溶液，振荡片刻。如呈墨绿色或蓝色，表示有噻吩存在。

正 丁 醇

用无水碳酸钾或无水硫酸钙进行干燥。过滤后，将滤液进行分馏，收集纯品。

氯 仿

普通氯仿含有1%的乙醇（它是作为稳定剂而加入的）。除去乙醇可用其1/2体积的水洗涤氯仿5～6次，然后用无水氯化钙干燥24h，进行蒸馏。纯品要放置于暗处，以免受光分解而形成光气，氯仿不能用金属钠干燥，否则会发生爆炸。

甲 苯

用无水氯化钙对甲苯进行干燥，过滤后加入少量金属钠片，再进行蒸馏，即得无水甲苯。普通甲苯可能含有少量甲基噻吩。除去甲基噻吩可将1000mL甲苯加入100mL浓硫酸，摇荡约30min（温度不要超过30℃）除去酸层，然后再分别用水、10%碳酸钠水溶液和水洗涤，以无水氯化钙干燥过夜，过滤后进行蒸馏，收集纯品。

乙 酸 乙 酯

市售乙酸乙酯通常含有微量水、乙醇和醋酸，用5%碳酸钠水溶液洗涤后，再用饱和氯化钙水溶液洗涤数次，以无水碳酸钾或无水硫酸镁进行干燥。过滤后，进行蒸馏，即得纯品。

呋 喃

用粒状氢氧化钠或氢氧化钾干燥过夜，然后进行蒸馏，即得无水呋喃。呋喃容易吸水，蒸馏时要注意防潮。

四 氢 呋 喃

四氢呋喃含水久贮后，可能含有过氧化物，检验方法是将四氢呋喃加入等体积的2%碘化钾溶液或淀粉溶液中，再加入几滴酸摇匀，若呈蓝或紫色，证明有过氧化物。一般加入硫酸亚铁溶液（6mL浓硫酸用100mL水稀释，加入60g硫酸亚铁）和100mL水充分摇匀，分出四氢呋喃。

无水四氢呋喃可用氢化铝锂在隔绝潮气下回流（一般1000mL用2～4g氢化铝锂），直至在处理过的四氢呋喃中加入钠丝和二苯酮，出现深蓝色的二苯酮钠，且加热回流蓝色不褪为止。在氮气保护下蒸出，备用。

二 噁 烷

普通二噁烷中含有少量乙醛，缩醛和水。在保存时缩醛水解产生乙醛，游离的乙醛会导致过氧化物的生成。通常用下列方法精制：

在500mL二噁烷中加入7mL浓盐酸和50mL水，在通风柜中加热回流12h，回流时缓慢地将氮气通入溶液以除去乙醛。待溶液冷却后加入粒状氢氧化钾直至不再溶解。分去水层,有机层中再加入粒状氢氧化钾振摇除去痕量水。将有机层放入干燥的圆底烧瓶中,加入金属钠加热回流10~12h,使金属钠最终保持光亮。最后蒸馏收集101℃馏分。

N,N-二甲基甲酰胺(DMF)

普通的 N,N-二甲基甲酰胺中含有少量的水、胺和甲醛等杂质。在常压蒸馏时有些分解产生二甲胺和一氧化碳。若有酸或碱存在时分解加快,如用固体氢氧化钾或氢氧化钠干燥数小时,会发生部分分解。因此它的提纯最好是用硫酸钙、硫酸镁、氧化钡、硅胶或分子筛干燥,然后减压蒸馏收集76℃/5.200kPa(39mmHg)的馏分。精制后的 N,N-二甲基甲酰胺最好放入分子筛后保存。

二 氯 甲 烷

使用二氯甲烷比氯仿安全,因此常常用它来代替氯仿作为比水重的萃取溶剂,普通的二氯甲烷一般都能直接作萃取剂使用。如需纯化,可用5%碳酸钠溶液洗涤,再用水洗涤,然后用无水氯化钙干燥,蒸馏收集40~41℃的馏分。

丙 酮

普通丙酮中常含有少量水及甲醇、乙醛等还原性杂质,分析纯的丙酮中即使有机杂质含量已少于0.1%,而水的含量仍达1%。它的纯化采用如下方法。

在500mL丙酮中加入2~3g高锰酸钾加热回流,以除去少量还原性杂质。若高锰酸钾紫色很快消失,则需再加入少量高锰酸钾继续回流,直至紫色不再消失为止。蒸出丙酮,然后用无水碳酸钾和无水碳酸钙干燥,蒸馏收集56~57℃馏分。

附录5 危险化学品使用知识

化学工作者经常使用各种各样的化学药品进行工作。常用化学药品的危险性,大体可分为易燃、易爆和有毒三类,现分述如下。

易燃化学药品

可燃气体:氢、乙胺、氯乙烷、乙烯、煤气、氢气、硫化氢、甲烷、氯甲烷、二氧化硫等。

易燃液体:汽油、乙醚、乙醛、二硫化碳、石油醚、苯、甲苯、二甲苯、丙酮、乙酸乙酯、甲醇、乙醇等。

易燃固体:红磷、三硫化二磷、萘、镁、铝粉等。黄磷为自燃固体。

可以看出,大部分有机溶剂均为易燃物质,若使用或保管不当,极易引起燃烧事故,故需特别注意。

易爆炸化学药品

气体混合物的反应速率随成分而异,当反应速率达到一定限度时,即会引起爆炸。

经常使用的乙醚,不但其蒸气能与空气或氧混合,形成爆炸混合物,放置过久的乙醚被氧化生成的过氧化物在蒸馏时也会引起爆炸。此外,四氢呋喃等环醚亦会产生过氧化物而引起爆炸。

某些以较高速度进行的放热反应，因生成大量气体也会引起爆炸并伴随着燃烧。

一般说来，易爆物质大多含有以下结构或官能团：

—O—O—	臭氧、过氧化物
—O—Cl—	氯酸盐、高氯酸盐
=N—Cl	氮的氯化物
—N=O	亚硝基化合物
—N=N—	重氮及叠氮化合物
—N=C	雷酸盐
—NO$_2$	硝基化合物（三硝基甲苯、苦味酸盐）
—C≡C—	乙炔化合物（乙炔金属盐）；

自行爆炸的有：高氯酸铵、硝酸铵、浓高氯酸、雷酸汞、三硝基甲苯等。

混合发生爆炸的有：

① 高氯酸＋酒精或其他有机物；

② 高锰酸钾＋甘油或其他有机物；

③ 高锰酸钾＋硫酸或硫；

④ 硝酸＋镁或碘化氢；

⑤ 硝酸铵＋酯类或其他有机物；

⑥ 硝酸＋锌粉＋水；

⑦ 硝酸盐＋氯化亚锡；

⑧ 过氧化物＋铝＋水；

⑨ 硫＋氧化汞；

⑩ 金属钠或钾＋水。

氧化物与有机物接触，极易引起爆炸。在使用浓硝酸、高氯酸及过氧化氮等时，必须特别注意。

防止爆炸除本书第一部分已叙述的知识外，还必须注意以下几点：

① 进行可能爆炸的实验，必须在特殊设计的防爆炸地方进行。使用可能发生爆炸的化学药品时，必须做好个人防护，戴面罩或防护眼镜，在不碎玻璃通风橱中进行操作；并设法减少药品用量或浓度，进行小量试验。对不了解性能的实验，切勿大意。

② 苦味酸需保存在水中，某些过氧化物（如过氧化苯甲酰）必须加水保存。

③ 易爆炸残渣必须妥善处理，不得任意乱丢。

有毒化学药品

日常接触的化学药品，有的是剧毒，使用时必须十分小心。有的药品长期接触或接触过多，也会引起急性或慢性中毒，影响健康。但只要掌握有毒化学药品的特性并且加以防护，就可避免或把中毒机会减少到最低程度。

有毒化学药品通常由下列途径侵入人体。

① 由呼吸道侵入。故有毒实验必须在通风橱内进行，并经常注意保持室内空气流畅。

② 由皮肤黏膜侵入。眼睛的角膜对化学药品非常敏感，故进行实验时，必须戴防护眼镜，进行实验操作时，注意勿使药品直接接触皮肤，手或皮肤有伤口时更须特别小心。

③ 由消化道侵入。这种情况不多。为防止中毒，任何药品不得用口尝味，严禁在实

室进食，实验结束后必须洗手。

常见的有毒化学药品如下。

① 有毒气体：氯、氟、氰氢酸、氟化氢、溴化氢、氯化氢、二氧化硫、硫化氢、光气、氨、一氧化碳等均为窒息性或具刺激性气体。在使用以上气体或进行有以上气体产生的实验时，必须在通风良好的通风橱中进行，并设法吸收有毒气体减少对环境的污染。如遇大量有害气体逸至室内，应立即关闭气体发生装置，迅速停止实验，关闭火源、电源，离开现场。如发生伤害事故，应视情况及时加以处理。

② 强酸和强碱：硝酸、硫酸、盐酸、氢氧化钠、氢氧化钾等均刺激皮肤，有腐蚀作用，造成化学烧伤。若吸入强酸烟雾，会刺激呼吸道，使用时应加倍小心，并严格按规定的操作进行。

③ 无机化学药品

氰化物及氰氢酸：毒性极强、致毒作用极快，空气中氰化氢含量达万分之三，数分钟内即可致人死亡，使用时须特别注意。氰化物必须密封保存，要有严格的领用保管制度，取用时必须戴口罩、防护眼镜及手套，手上有伤口时不得进行使用氰化物的实验。研碎氰化物时，必须用有盖研钵，在通风橱进行（不抽风）；使用过的仪器、桌面均得亲自收拾，用水冲净，手及脸亦应仔细洗净；实验服可能污染，必须及时换洗。

汞：室温下即能蒸发，毒性极强，能导致急性或慢性中毒。使用时必须注意室内通风；提纯或处理，必须在通风橱内进行。如果泼翻，可用水泵减压收集，尽可能收集完全。无法收集的细粒，可用硫黄粉、锌粉或三氯化铁溶液清除。

溴：液溴可致皮肤烧伤，蒸气刺激黏膜，甚至可使眼睛失明。使用时必须在通风橱中进行。盛溴的玻璃瓶须密闭后放在金属罐中，妥为存放，以免撞倒或打翻；如泼翻或打破，应立即用沙掩盖。如皮肤灼伤立即用稀乙醇冲洗或用大量甘油按摩，然后涂硼酸凡士林。

金属钠、钾：遇水即发生燃烧、爆炸，使用时须小心。钠、钾应保存在液体石蜡或煤油中，装入铁罐中盖好，放在干燥处。

④ 有机化学药品

有机溶剂：有机溶剂均为脂溶性液体，对皮肤黏膜有刺激作用，对神经系统有选择作用。如苯，不但刺激皮肤，易引起顽固湿疹，对造血系统及中枢神经系统均有严重损害。再如甲醇对视神经特别有害。在条件许可情况下最好用毒性较低的石油醚、醚、丙酮、甲苯、二甲苯代替二硫化碳、苯和卤代烷类。

硫酸二甲酯：鼻吸入及皮肤吸收均可中毒，且有潜伏期，中毒后感到呼吸道灼痛，对中枢神经影响大，滴在皮肤上能引起坏死、溃疡，恢复慢。

芳香硝基化合物：化合物所含硝基愈多毒性愈大，在硝基化合物中增加氯原子，亦将增加毒性。此类化合物的特点是能迅速被皮肤吸收，中毒后引起顽固性贫血及黄疸病，刺激皮肤引起湿疹。

苯酚：能够灼伤皮肤，引起坏死或皮炎，沾染后应立即用温水及稀酒精洗。

生物碱：大多数生物碱具强烈毒性，皮肤亦可吸收，少量可导致危险中毒甚至死亡。

致癌物：很多烷基化剂长期摄入体内有致癌作用，应予注意。其中包括硫酸二甲酯、对甲苯磺酸甲酯、N-甲基-N-亚硝基脲素、亚硝基二甲胺、偶氮乙烷以及一些丙烯酯类等。一些芳香胺类，由于在肝脏中经代谢而生成 N-羟基化合物而具有致癌作用，其中包括 2-乙酰氨基芴、4-乙酰氨基联苯、2-乙酰氨基苯酚、2-萘胺、4-二甲氨基偶氮苯等。部分稠环芳香烃化合物，如3,4-苯并蒽、1,2,5,6-二苯并蒽和9-及10-甲基-1,2-苯并蒽等，都是致癌物，而9,10-二甲基-1,2-苯并蒽则属于强致癌物。

使用有毒药品时必须小心，了解其性质与使用方法。不要沾污皮肤、吸入蒸气及溅入口中。最好在通风橱内操作，必要时戴防护眼镜及手套，小心开启瓶塞以免破损散出。使用过的仪器，应亲自冲洗干净，残渣废料丢在废物缸内。经常保持实验室及台面整洁，也是避免发生事故的重要措施。实验结束后必须养成洗手的习惯。

附录6　典型的红外光谱吸收谱带

2.5～3.1μm 区域	波长/μm	波数/cm^{-1}
O—H 伸缩		
醇类和酚类		
自由 O—H	2.75～2.8(尖)	3636～3571
氢键		
二聚体	2.80～2.90(尖)	3571～3448
多聚体	2.94～3.10(宽)	3401～3226
羧酸	3.0～4.0(宽)	3333～2500
N—H 伸缩		
胺和酰胺		
自由 N—H		
伯	2.8～2.9 和 2.9～3.0	3571～3448 和 3448～3333
仲	2.85～2.95	3509～3390
氢键	约 2.99～3.25(宽)	3344～3077
乙炔型 C—H	3.0～3.1(尖)	3333～3226
3.2～4.0μm 区域	波长/μm	波数/cm^{-1}
C—H 伸缩		
烯类	3.23～3.32	3096～3012
芳烃类	约 3.30	约 3030
烷类	3.37～3.52	2967～2841
		2899～2817
醛 $\left(\begin{array}{c}O\\ \|\\ -C-H\end{array}\right)$	3.45～3.55 和 3.6～3.7	2899～2817 和 2778～2703
O—甲基，N—甲基	3.54～3.62	2825～2762
4.2～5.0μm 区域	波长/μm	波数/cm^{-1}
C≡Y 伸缩		
腈（—C≡N）	4.42～4.51	2262～2217
乙炔（—C≡C—）	4.43～4.76	2257～2101
5.5～6.0μm 区域	波长/μm	波数/cm^{-1}
C=O 伸缩		
酸酐 $\left(\begin{array}{c}O\quad\;O\\ \|\quad\;\|\\ -C-O-C-\end{array}\right)$ 酰氯	约 5.5 和约 5.7	约 1818 和约 1754
脂肪酰氯 $\left(\begin{array}{c}O\\ \|\\ R-C-Cl\end{array}\right)$	约 5.55	约 1802

续表

5.5～6.0μm 区域	波长/μm	波数/cm^{-1}
芳香酰氯酯 ($Ar-\overset{O}{\underset{\|}{C}}-Cl$)	约 5.68	约 1761
饱和酯 ($R-\overset{O}{\underset{\|}{C}}-OR$)	5.71～5.78	1751～1730
α,β-不饱和酯和芳香酸酯	5.78～5.82	1730～1718
饱和酸 ($R-\overset{O}{\underset{\|}{C}}-OH$)	5.80～5.88	1724～1701
α,β-不饱和酸和芳香酸	5.88～5.94	1701～1684
饱和醛（RCH=O）	5.75～5.81	1739～1721
α,β-不饱和醛 ($\overset{\|}{C}=\overset{\|}{C}-CH=O$)	5.78～5.93	1730～1686
芳香醛（Ar—CH=O）酮	5.83～5.89	1715～1698
环戊酮	约 5.75	约 1739
饱和酮 ($R-\overset{O}{\underset{\|}{C}}-R$)	约 5.83	约 1715
芳香酮 ($R-\overset{O}{\underset{\|}{C}}-Ar$)	5.88～5.95	1701～1681
α,β-不饱和酮 ($-\overset{\|}{C}=\overset{\|}{C}-\overset{O}{\underset{\|}{C}}-R$)	5.94～6.01	1684～1664
酰胺 ($-\overset{O}{\underset{\|}{C}}-N\overset{}{\underset{}{<}}$)	5.88～6.14	1701～1629
6.0～7.4μm 区域	**波长/μm**	**波数/cm^{-1}**
C=C 伸缩		
单烯	5.95～6.17	1681～1621
二烯（C=C—C=C）	约 6.06 和约 6.25	约 1650 和约 1600
芳环	6.25～6.35	1600～1575
	6.6～6.7 和 6.9	1515～1493 和 1450
硝基伸缩		
硝基烷类（RNO$_2$）	约 6.4 和 7.4	约 1563 和约 1351
硝基芳烃（ArNO$_2$）	约 6.6 和 7.4	约 1515 和约 1351
—NH$_2$ 弯曲	6.1～6.3	1639～1587
—CH$_2$—变形	6.74～6.92	1485～1445
C—CH$_3$ 变形	6.8～7.0 和 7.25～7.32	1471～1429 和 1379～1366
7.4～11μm 区域（指纹区）	**波长/μm**	**波数/cm^{-1}**
C—O 伸缩		
酚（Ar—OH）	约 8.33	约 1200
醇		
伯（RCH$_2$OH）	约 9.5	约 1053

续表

指纹区 7.4~11μm 区域	波长/μm	波数/cm^{-1}
仲(R_2CHOH)	约 9.1	约 1099
叔(R_3COH)	约 8.7	约 1149
醚(R—O—R)	8.7~9.35	1150~1070
C—N 伸缩		
胺		
脂肪(R—N<)	8.2~9.8	1220~1020
芳香(Ar—N<)	7.4~8.0	1350~1250

附录 7 一些常见基团质子的化学位移值

常见基团质子	化学位移(δ)	常见基团质子	化学位移(δ)
RCH$_3$	0.9	Ar—CH$_3$	2.3
R$_2$CH$_2$	1.3	R—CO—CH$_3$	2.2
R$_3$CH	1.5		
RCH$_2$Cl	3.5~4.0	R—CO—O—CH$_3$	3.6
RCH$_2$Br	3.0~3.7		
RCH$_2$I	2.0~3.5	R—O—H	3.0~6.0
R—O—CH$_3$	3.2~3.5	Ar—O—H	6.0~8.0
R—O—CH$_2$—CH$_3$	1.2~1.4	R—CHO	9.0~10.0
R—O—(CH$_2$)$_2$CH$_3$	0.9~1.1		
C=C—H	5.0~5.3	R—CO—O—H	10.5~11.5
C≡C—H	2.5	R—NH$_2$	1.0~4.0
Ar—H	6.5~8.0	Ar—NH$_2$	3.0~4.5
C=C—CH$_3$	1.7	R$_2$N—CH$_3$	2.2
C≡C—CH$_3$	1.8		

附录 8 实验单元操作一览表

实验名称	熔点测定	重结晶水	重结晶有机溶剂	无水反应	搅拌	回流	滴液	分液	干燥	蒸馏	分水器	水蒸气蒸馏	减压蒸馏	薄层色谱	柱色谱	过滤
环己烯								✓	✓	✓						
2-methyl-2-butene								✓	✓	✓						
溴乙烷								✓	✓	✓						
溴丁烷						✓		✓	✓	✓						
α,β-二溴苯乙烷			✓			✓										✓
t-butyl chloride																
苯甲醇						✓	✓	✓								
二苯甲醇	✓		✓						✓							✓
triphenylmethanol	✓				✓	✓	✓	✓	✓							✓

续表

实验名称	熔点测定	重结晶水	重结晶有机溶剂	无水反应	搅拌	回流	滴液	分液	干燥	蒸馏	分水器	水蒸气蒸馏	减压蒸馏	薄层色谱	柱色谱	过滤
乙醚							✓	✓	✓	✓						
正丁醚						✓		✓	✓	✓	✓					
4-nitroanisol	✓		✓							✓						✓
环己酮							✓	✓	✓	✓						
苯乙酮					✓	✓	✓	✓	✓				✓			✓
茴香基丙酮					✓	✓	✓	✓	✓				✓			
benzophenone					✓	✓	✓	✓	✓				✓			
diphenylethanedione						✓	✓	✓	✓	✓			✓			
苯甲酸	✓	✓				✓	✓		✓							✓
对硝基苯甲酸	✓		✓			✓	✓		✓							✓
己二酸	✓	✓				✓	✓		✓							✓
香豆素-3-羧酸	✓					✓	✓		✓							✓
cinnamic acid	✓					✓	✓		✓							✓
邻苯二甲酸二丁酯					✓		✓	✓	✓		✓		✓			
丁二酸酐	✓					✓										✓
苯甲酸乙酯					✓			✓		✓						✓
巯基乙酸铵					✓		✓									
ethyl acetate						✓		✓	✓	✓						
acetanilide	✓	✓				✓			✓							✓
硝基苯																
间二硝基苯	✓		✓		✓	✓			✓							✓
2-nitroresorcinol	✓		✓		✓								✓			✓
间硝基苯胺	✓				✓	✓										✓
邻氨基苯甲酸						✓	✓									✓
溴化四丁基铵					✓			✓								✓
溴化四乙基铵					✓			✓								✓
benzyltriethylammonium chloride					✓			✓								✓
氯苯					✓		✓	✓	✓	✓		✓				✓
甲基红	✓	✓			✓				✓							✓
Ⅱ号橙	✓	✓			✓				✓							✓
methyl orange	✓	✓			✓				✓							✓
双酚-S				✓						✓						✓
p-toluene sulfonic acid	✓	✓				✓			✓		✓					✓
双酚 A			✓			✓										✓
安息香					✓	✓	✓						✓			
ethyl acetoacetate				✓	✓			✓	✓	✓						✓
呋喃甲醇和呋喃甲酸	✓	✓				✓			✓	✓						✓
benzoic alcohol and benzoic acid	✓	✓			✓			✓	✓	✓						✓

续表

实验名称	熔点测定	重结晶 水	重结晶 有机溶剂	无水反应	搅拌	回流	滴液	分液	干燥	蒸馏	分水器	水蒸气蒸馏	减压蒸馏	薄层色谱	柱色谱	过滤
苯甲酸	✓	✓		✓	✓	✓	✓	✓								✓
2-甲基-2-己醇				✓	✓	✓	✓	✓		✓						
malachite green and crystal violet					✓		✓									✓
氢化肉桂酸	✓				✓				✓				✓			
cis-4-t-butylcyclohexanol									✓							✓
camphor	✓													✓		
绿色植物色素的提取					✓	✓								✓	✓	
番茄红素、β-胡萝卜素的提取					✓									✓	✓	
果胶的提取					✓											✓
extraction of caffeine from tea	✓					✓				✓						✓
苯乙胺的制备及拆分						✓		✓		✓			✓			
扁桃酸		✓	✓		✓	✓	✓	✓							✓	
1-苯基-2,2-二氯环丙烷				✓		✓	✓									
7,7-dichlorobicyclo[4.1.0]heptane					✓	✓	✓									
五乙酸葡萄糖酯			✓	✓												✓
酪蛋白和乳糖					✓											✓
乙酰淀粉					✓											✓
L-酪氨酸甲酯							✓						✓			
3-氨基邻苯二甲酰肼																✓
二苯甲酮																✓
肉桂酸	✓	✓														✓
3-烯丙基-2,4-戊二酮													✓			
methy benzoate						✓		✓	✓							
1,2,3,4-tetrahydrocarbazole				✓												✓
二茂铁			✓		✓	✓	✓			✓			✓			
丁基锂				✓	✓		✓									
acetyl ferrocene														✓	✓	✓
局部麻醉剂苯佐卡因			✓		✓	✓			✓							
磺胺药物-磺胺			✓		✓											
解热镇痛药-非那西汀		✓			✓											
止咳酮				✓	✓				✓				✓			
5,5-二苯基乙内酰脲			✓		✓											
aspirin			✓													
杀灭菊酯					✓			✓	✓	✓						
敌敌畏					✓		✓	✓	✓							
N,N-diethyl-m-toluamide				✓	✓	✓								✓		

参 考 文 献

[1] 周宁怀，王德琳主编. 微型有机化学实验. 北京：科学出版社，1999.
[2] 武汉大学化学与分子科学学院实验中心编. 有机化学实验. 武汉：武汉大学出版社，2004.
[3] 刘玉英，马晨主编. 微型有机化学实验. 济南：山东大学出版社，1999.
[4] 顾可权等编. 半微量有机制备. 北京：高等教育出版社，1990.
[5] Thomas L Jacobs, William E Truce, G Ross Robeertson. Laboratory Practice of Organic Chemistry. 5th ed.. New York：MacMillan Publishing Co. Inc, 1974.
[6] 兰州大学，复旦大学化学系有机化学教研室编. 有机化学实验. 第2版. 北京：高等教育出版社，1994.
[7] 北京大学化学系有机化学教研室编. 有机化学实验. 北京：北京大学出版社，1990.
[8] 周科衍，高占先主编. 有机化学实验. 第3版. 北京：高等教育出版社，1996.
[9] 奚关根，赵长宏，赵中德编著. 有机化学实验. 上海：华东理工大学出版社，1995.
[10] 鲍慈光，赵逸云，侯秀英等编. 应用化学实验. 北京：科学出版社，1996.
[11] John W Lehman. Operational Organic Chemistry A Laboratory Course. 2nd. ed.. Massachusetts：Allyn and Bacon Inc, 1988.
[12] Pavia D L, Lampman Gary M, Kriz Jr George S. Introduction to Organic Laboratory Techniques. 3rd ed. Philadephia：Saunders College Publishing, 1988.
[13] 陆涛，陈继俊主编. 有机化学实验与指导. 第2版. 北京：中国医药科技出版社，2006.
[14] G D Christian, J E O'Reilly 主编. 仪器分析. 王镇浦，王镇棣译. 北京：北京大学出版社，1991.
[15] Clark F Most Jr. Experimental Organic Chemistry. New York：John Wiley & Sons, 1988.
[16] 韩广甸等编译. 有机制备化学手册. 北京：石油化学工业出版社，1977.
[17] Vogel's Text book of Practical Organic Chemistry. 5th Ed. Prentice Hall, 1996.